Chemistry 101
Lecture Notes

Dana S. Chatellier

University of Delaware

KENDALL/HUNT PUBLISHING COMPANY
4050 Westmark Drive   Dubuque, Iowa 52002

DEDICATION....

....is required in order to succeed in anything you attempt -- learning about chemistry, becoming an athlete or an artist, or just leading a balanced life.

This book is dedicated to a lot of dedicated people....

....my family, immediate and extended, whose dedication to my welfare has made me who I am, and whose dedication to setting good examples for me has inspired me to do the same for others....

....my teachers and colleagues, former and current, whose dedication to shaping and stimulating young minds has provided me with many role models, and whose dedication to the welfare of the <u>whole</u> person (not just the student) has inspired me to do the same for others....

....my Christian friends, near and far, whose dedication to the service of the Lord has taught me what love is, and whose dedication to the welfare of our fellow human beings has inspired me to do the same for others....

....but mostly, my students, past and present, whose dedication to tasks which they often find unpleasant has kept me going when times got tough, and whose dedication to doing the best that they can do has inspired me to do the same for other, future students.

This book is <u>your</u> book, for you are all a part of me.

--Dana S. Chatellier
February, 1989

TABLE OF CONTENTS

                                    PREFACE

        This book is a typewritten version of the lecture notes which I have
used for the last several years in teaching Chemistry 101 at the University of
Delaware. Chemistry 101 is the first-semester course of a two-semester sequence
in chemistry intended largely for non-science majors. Most general chemistry
textbooks fall into one of two categories: those intended for a one-semester
course for non-science majors, or those intended for a two-semester sequence for
science majors. As the demands of Chemistry 101 (and its second-semester
counterpart, Chemistry 102) fall somewhere between these two extremes, it was
deemed useful (if not necessary!) for a good set of lecture notes to accompany the
textbook for the course, the better to assist the student in approaching the study
of chemistry for what may be the first time.

        My original lecture notes were compiled during my first teaching
position, at Willamette University in Salem, Oregon. They were adapted from the
text in use at that time at that institution -- "Fundamentals of Chemistry" by
James E. Brady and John R. Holum. Since then, they have undergone considerable
revision and alteration. In preparing this manuscript, I have attempted to make
the material therein adaptable to many of the existing general chemistry textbooks,
including the one currently in use in Chemistry 101 at the University of Delaware
-- namely, "General Chemistry" by Henry F. Holtzclaw, Jr. and William R. Robinson.
The order in which topics are presented is similar to that in most current general
chemistry textbooks. However, the modular format employed -- that is, one topic
is covered per page in most instances -- allows for facile rearrangement of topics
simply by re-ordering pages.

        This book is not a textbook. Indeed, it would be difficult to use it
as such -- there are very few illustrations, and not very many problems for
students to practice working. This was done intentionally, since this book is
intended as a supplement to a general chemistry textbook, most of which have
illustrations and practice problems galore. The advantage to the student of
having this book is that it is written in a style which is more conversational than
that found in most textbooks -- in short, it's written the way I teach it! Many
of my students have found this feature to be very beneficial. In addition, this
book allows the student to simply listen to my lectures without the bother of
taking notes, if such is the student's desire. For some students, this may be the
best way to learn!

        This book will obviously find its greatest applicability in courses in
general chemistry which I teach, but it is hoped that students in other courses
(and possibly at other institutions) will find it useful as well. Chemistry is a
fascinating field of study, and one whose importance in today's technological world
cannot be denied. If this book helps people understand the world and the age in
which they live, then it will have served its purpose.

                                                    --Dana S. Chatellier
                                                      February, 1989

TO THE STUDENT:

This is <u>your</u> book. You've bought it and paid for it -- it's yours.
Use it that way! Make notes to yourself in the margins or between lines when you
feel the need. (That's why it was typed double-spaced!) Draw pictures to
illustrate an example or a chemical demonstration done in class. (This book is
almost pictureless, since there are many good illustrations in your textbook.)
Tear out pages and rearrange them when necessary. Throw the book at the wall (not
at your roommate!) when studying chemistry gets you frustrated! This is <u>your</u> book,
so use it however <u>you</u> see fit.

Good luck! The study of chemistry is challenging, but most students
find that they can master it with enough hard work and patience. Read the textbook
and these lecture notes, work practice problems, and see your instructor if you're
having problems. (Most students don't admit that they're having problems until
they do poorly on a few exams. That's too late! See your instructors <u>before</u> you
run into serious problems -- it will make for a much more enjoyable experience for
everyone.)

Best wishes for a rewarding semester learning about chemistry.

--Dana S. Chatellier
February, 1989

SCIENCE AND THE SCIENTIFIC METHOD:

Science is a way of trying to understand the universe in which we live. The key word here is "understand". In the words of author Robert Heinlein, "The difference between science and the fuzzy subjects is that science requires reasoning, while those other subjects merely require scholarship."[1]

Scientists, according to mathematician John von Neumann, "mainly make models. A model is a mathematical construct which describes observed phenomena. The justification of such a construct is that it is expected to work."[2] Adds Leo Kadanoff, "One is surprised that a construct of one's own mind can actually be realized in the honest-to-goodness world out there. A great shock, a great joy."[2]

von Neumann's comments are a brief way of summarizing the <u>scientific method</u>, which is the scientist's way of trying to understand the universe. The scientific method involves choosing a natural phenomenon to be studied, doing <u>experiments</u> to see how the object being studied interacts with other things, and recording the results of these experiments as <u>data</u>. Data may include both qualitative observations and quantitative measurements. If any patterns or trends are observed in a collection of data, and if further experiments confirm that these trends are broadly applicable, then the trends are referred to as <u>laws</u>. Laws tell <u>what</u> happens, but they say nothing about <u>why</u> it happens. Scientists form models called <u>hypotheses</u> to try to explain <u>why</u> things happen. A hypothesis whose predictions are confirmed over a long period of time is called a <u>theory</u>. Theories and laws are <u>never</u> proven, but they can be <u>disproven</u>. "No amount of experimentation can ever prove me right; a single experiment can prove me wrong." --A. Einstein[3]

---

1) Heinlein, R. A., <u>Time Enough For Love</u>. G. P. Putnam's Sons, 1973.

2) Quoted by James Gleick in <u>Chaos: Making a New Science</u>. Penguin Books, 1987.

3) Quoted by Dr. Laurence J. Peter in <u>Peter's Quotations</u>. William Morrow and Company, Inc., 1977.

CHEMISTRY -- THE CENTRAL SCIENCE:

In the words of Dr. John A. Conkling, Executive Director of the American Pyrotechnics Association, "Chemistry is a science concerned with substances and their properties, with the changes or reactions whereby other substances are formed, with the conditions necessary for bringing about or preventing these changes, and with the relative amounts of matter and energy involved."[1] Chemistry has been called "the central science" because it is so closely related to other sciences. For example, among the many branches of chemistry are biochemistry (the chemistry of living things), organic chemistry (the chemistry of the substances of which living things are made), and physical chemistry (chemistry related to physics, involving studies of matter and energy).

Matter is the "stuff" of which any physical object is made -- anything that has mass (or weight) and occupies space is made of matter. Matter has many properties (or characteristics) which can be measured. The properties of matter can be classified as either extensive properties (which depend on the amount of matter present) or intensive properties (which don't depend on the amount of matter present). Extensive properties include the mass and volume of an object; intensive properties include the object's melting point and boiling point. Note that an object's density (an intensive property) is just the ratio of its mass to its volume (two extensive properties): Density = mass/volume.

The properties of matter can also be classified as chemical properties (which depend on the substance's interaction with other substances) or as physical properties (which don't depend on the substance's interaction with any other substance). The properties mentioned above are all physical properties. Examples of chemical properties include a substance's acidity, flammability, and toxicity.

---

1) From a lecture given at Washington College, Chestertown, Maryland, September, 1974.

CLASSIFICATIONS OF MATTER:

Most of the matter which is found in nature is a <u>mixture</u> of two or more substances. <u>Impure substances</u> have <u>variable compositions</u> -- that is, they can be mixed in <u>any proportion</u> that we choose. As an example, consider <u>coffee</u> -- a mixture of water, caffeine, sugar, cream, etc. These ingredients can be mixed in whatever proportions are desired. This is fortunate, since some people like their coffee stronger (or sweeter, or blacker) than others.

Impure substances occur in two varieties: they are either <u>homogeneous mixtures</u> or <u>heterogeneous mixtures</u>. <u>Heterogeneous</u> mixtures are composed of at least two distinct <u>phases</u>. <u>Phases</u> are regions whose physical and chemical properties differ from each other, but are <u>constant</u> within that region. As an example, consider a mixture of oil and water. "Oil and water don't mix" -- they separate into two <u>layers</u>: an oil layer (oil phase) and a water layer (water phase). Mixtures which contain only <u>one</u> phase are called <u>homogeneous</u> mixtures. As an example, consider the cup of coffee again. Coffee, if stirred properly, does not separate into a water layer, a cream layer, a caffeine layer, and so on. There's just <u>one</u> phase (the "coffee phase") in a homogeneous mixture.

Much of the hard work that many chemists do involves separating mixtures into <u>pure substances</u>, which have <u>constant compositions</u>. Pure substances fall into one of two classifications: <u>elements</u> and <u>compounds</u>. <u>Elements</u> are the simplest forms of matter. They cannot be broken down into simpler substances by normal chemical methods. The smallest particles into which an element can be divided without losing its chemical properties are called <u>atoms</u>. Each element is composed of only <u>one</u> kind of atom. <u>Compounds</u> are formed from two or more elements combined chemically in a <u>fixed ratio</u>. Many compounds are made of <u>molecules</u>, which are small particles composed of two or more atoms.

THE CHEMIST'S "SHORTHAND":

When chemists write about the elements and compounds that they use, they often use a "shorthand" notation to be able to avoid writing the full names of the substances involved. Learn this "shorthand" notation -- it saves time!

Elements are written using chemical symbols. Chemical symbols are simply one-letter or two-letter abbreviations for the names of the elements. One-letter symbols are used for many common elements, such as hydrogen (H), carbon (C), and nitrogen (N). A second letter is added to distinguish between two elements whose names begin with the same letter. Elements with two-letter symbols include helium (He), chlorine (Cl), and niobium (Nb). A few elements have symbols which are derived from their Latin names. Examples include mercury (Hg, from the Latin hydrargyrum), copper (Cu, from cuprum), and sodium (Na, from natrium). You should learn the symbols and names of the more common elements as soon as you can.

Compounds are written using chemical formulas. Water is a compound whose formula is the well-known H_2O. This formula means that one molecule of water is made of two atoms of hydrogen (H) and one atom of oxygen (O) which are held together in some way. Many compounds are made of more complex molecules. A common example is sucrose (table sugar), which has the formula $C_{12}H_{22}O_{11}$. This formula indicates that twelve carbon atoms, twenty-two hydrogen atoms, and eleven oxygen atoms are joined to each other to form one molecule of sucrose.

Hydrates are compounds whose crystals contain loosely-held molecules of water in their crystal structures. Their formulas are written so as to clearly indicate this. For example, the formula $CoCl_2 \cdot 6H_2O$ shows that six molecules of water are connected to each $CoCl_2$ unit. (Note the importance of the correct use of capital letters when writing chemical symbols and formulas. "$CoCl_2$" is a crystalline solid; "$COCl_2$" is a poisonous gas. Little things mean a lot!)

THE METRIC SYSTEM:

By international agreement, scientists use the metric system to measure the various quantities that they need to know in their laboratories. The metric system is also called the SI system. SI comes from Systeme Internationale d'Unites -- the International System of Units. Most industrialized countries (with the notable exception of the United States!) use the metric system for everyday transactions.

One of the advantages of the metric system is that it is based on powers of ten, which makes it easy to convert from one unit to another. In the metric system, certain prefixes are used to indicate multiples or fractions of the base unit. For example, "kilo-" means one thousand of the base unit (1000 grams = 1 kilogram), "centi-" means one one-hundredth of the base unit (100 centimeters = 1 meter), and "milli-" means one one-thousandth of the base unit (1000 milliliters = 1 liter). By contrast, there are two pints in a quart, three feet in a yard, four quarts in a gallon, twelve inches in a foot, etc. Use metric -- it's easier!

Conversion to the metric system has met with resistance in the United States, presumably because people are unfamiliar with the ways in which metric units compare with other units. A table of metric conversion factors is provided.

Measured Quantity	SI Unit of Measure	Conversion Factor Equations
Mass	Kilogram	1 kg = 2.205 pounds
	Gram	28.35 g = 1 ounce
Distance	Meter	1 m = 39.37 inches
	Kilometer	1.609 km = 1 mile
Volume	Cubic Meter	1 m^3 = 1000 liters
	Liter	1 L = 1.057 quarts

MEASURING TEMPERATURE:

In the United States, temperature is measured using the _Fahrenheit_ scale, but in most other countries, the _Celsius_ (sometimes called _centigrade_) scale is used. Scientists use the Celsius scale and another scale called the _Kelvin_ scale in the laboratory. These two temperature scales are well-suited for the kinds of measurements and calculations that scientists typically make.

On the Fahrenheit scale, the freezing point of water is 32 ^{O}F, and the boiling point of water is 212 ^{O}F. The Celsius scale is sometimes called the _centigrade_ scale because these same two physical phenomena, the freezing and boiling of water, are separated by _one hundred_ degrees ("centi-" means 1/100.). The freezing point of water is 0 ^{O}C, and the boiling point of water is 100 ^{O}C. The same two points are also separated by one hundred _Kelvins_ (Note: _not_ "degrees Kelvin"!): water freezes at 273 K, and water boils at 373 K. _Zero_ on the Kelvin scale corresponds to _absolute zero_, the coldest possible temperature. This equals -459 ^{O}F and -273 ^{O}C, but there are no "below zero" temperatures on the Kelvin scale. The Kelvin scale is sometimes called the _absolute scale_ for this reason.

Converting a temperature from Celsius to Kelvin is easy: just add 273! Similarly, just subtract 273 from a Kelvin temperature to get the corresponding Celsius temperature. Conversion between the Fahrenheit and Celsius scales is a little bit trickier, but here's a relatively simple way to do it:[1] _Step 1_ -- add 40. _Step 2_ -- multiply by 9/5 if converting Celsius to Fahrenheit. Multiply by 5/9 if converting Fahrenheit to Celsius. _Step 3_ -- subtract 40.

Summary: K = ^{O}C + 273. ^{O}F = [(40 + ^{O}C)(9/5)] - 40 = (1.8)(^{O}C) + 32

^{O}C = K - 273. ^{O}C = [(40 + ^{O}F)(5/9)] - 40 = $\dfrac{(^O\text{F} - 32)}{1.8}$

------

1) The author wishes to thank Dr. John Burmeister, University of Delaware, for sharing this method.

$$\text{USING CONVERSION FACTORS:}$$

To convert 300 feet into yards, we need only to divide by 3, since 3 feet = 1 yard. (Everyone knows that!) But how would you approach such a problem if you <u>didn't</u> know simply to divide by 3 -- that is, if you were using units which aren't as familiar as feet and yards?

A <u>conversion factor</u> is simply an algebraic expression that equals <u>one</u>. As an example, let's consider the equation above: 3 feet = 1 yard. Since this is a true statement, then the following two equations must also be true:

$$\frac{3 \text{ feet}}{1 \text{ yard}} = 1 \qquad\qquad \frac{1 \text{ yard}}{3 \text{ feet}} = 1$$

To convert 300 feet into yards, all we need to do is multiply by <u>one</u> in the form of one of the above conversion factors. But which one shall we use? If we choose the conversion factor on the left (above), we get this result:

$$300 \text{ feet} \times \frac{3 \text{ feet}}{1 \text{ yard}} = 900\ \frac{\text{feet} \times \text{feet}}{\text{yards}} = 900 \text{ feet}^2/\text{yard}.$$

This is clearly <u>incorrect</u>, since we know that our final answer should be in <u>yards</u>, not some silly units like feet^2/yard! Using the other conversion factor above gives the correct result, <u>complete with the correct units</u>:

$$300 \text{ feet} \times \frac{1 \text{ yard}}{3 \text{ feet}} = 100\ \frac{\text{feet} \times \text{yards}}{\text{feet}} = 100 \text{ yards}.$$

The general method is to set up the conversion factors so that any <u>undesired</u> units will appear in both the <u>numerator</u> and the <u>denominator</u> of the final expression. These units will <u>cancel out</u>, leaving only the <u>desired</u> units.

<u>Problem</u>: A European car gets 12.5 kilometers per liter of gasoline. What is the mileage in miles per gallon? 1 mile = 1.609 km. 1 gallon = 3.786 L.

<u>Solution</u>: Setting up the necessary conversion factors, we get:

$$12.5\ \frac{\text{km}}{\text{L}} \times \frac{3.786 \text{ L}}{1 \text{ gallon}} \times \frac{1 \text{ mile}}{1.609 \text{ km}} = 29.4\ \frac{\text{miles}}{\text{gallon}}.$$

$$\text{SCIENTIFIC NOTATION:}$$

Problem: Light travels through space at a speed of 300,000,000 meters per second. Express this velocity in <u>furlongs per fortnight</u>!

Conversion Equations: 8 furlongs = 1 mile. 14 days = 1 fortnight.

Solution: Set up the necessary conversion factors!

$$300,000,000 \ \frac{m}{sec} \times \frac{60 \text{ sec}}{1 \text{ min}} \times \frac{60 \text{ min}}{1 \text{ hr}} \times \frac{24 \text{ hr}}{1 \text{ day}} \times \frac{14 \text{ days}}{1 \text{ fortnight}}$$

$$\times \frac{1 \text{ km}}{1000 \text{ m}} \times \frac{1 \text{ mile}}{1.609 \text{ km}} \times \frac{8 \text{ furlongs}}{1 \text{ mile}} = 1,800,000,000,000 \ \frac{furlongs}{fortnight} .$$

Scientists often deal with large numbers such as 300,000,000 and 1,800,000,000,000. They also need to use small numbers such as 0.000000656. <u>Scientific notation</u> has been developed to enable scientists to work with very large and very small numbers conveniently (and to save a lot of time and effort which would otherwise be spent writing a lot of zeroes!). To write a number in scientific notation, simply move the decimal point until the number that is formed is between <u>one</u> (1) and <u>ten</u> (10). Then, write "x 10^n" after the new number. The value of n is simply the number of decimal places that the decimal point would have to be moved to return to the original number. Some examples are provided.

Problem: Write 300,000,000 in scientific notation.

Solution: Another way to write 300,000,000 is 300000000. Notice that the decimal point occurs after the last zero. Moving this decimal point eight places to the <u>left</u> gives us the number <u>3</u>. Therefore, the value of n is <u>8</u>, and 300,000,000 would be written as 3 x 10^8 in scientific notation.

Problem: Write 0.000000656 in scientific notation.

Solution: Moving the decimal point seven places to the <u>right</u> gives us a number between one and ten -- namely, <u>6.56</u>. Therefore, the value of n is <u>-7</u>, and 0.000000656 would be written as 6.56 x 10^{-7} in scientific notation.

All measurements contain uncertainties. Surveys and opinion polls are usually reported as having a certain "margin for error". When you measure something (using a ruler or thermometer, for example), you can only read the number to a certain degree of precision -- your eyes work well enough to allow you to "read between the lines" to a certain extent, but not much more than that.

When a scientist measures something, s/he records not only the numerical value of the measurement, but also an indication of the uncertainty of the measurement. Suppose, for example, a mass was determined using a balance which can be read to the nearest centigram (0.01 gram). If the measured mass was 6.03 grams, the entry in the scientist's laboratory notebook would look something like this: "Mass of liquid = 6.03 ± 0.01 grams". (The "±" symbol is read as "plus or minus", implying an uncertainty of 0.01 grams.) For the scientist to record any digits beyond the "3" in the hundredths place would be meaningless, since there is an uncertainty in that place. If the hundredths place contains an uncertainty, the thousandths place and others beyond it must also be uncertain.

If the mass of the liquid had been recorded as simply "6.03 grams", an uncertainty in the hundredths place would be assumed. This is referred to as the significant figure convention -- the significant digits are all of those which are known with certainty, plus one digit which is uncertain. In the above case, the "6" and the "0" are known with certainty, and the "3" is uncertain. Hence, the measured number "6.03" is said to have three significant figures ("sig figs").

If you are not sure how many "sig figs" are present in a measured number, write the number in scientific notation! For example, the number 0.082060 may appear to have seven "sig figs", but writing it as 8.2060×10^{-2} shows that only five "sig figs" are present -- the first two zeroes are just "place holders".

USING SIGNIFICANT FIGURES IN CALCULATIONS:

Problem: If a sample of a liquid has a mass of 6.03 grams and a volume of 7.12 milliliters, what is the density of the liquid?

Solution: Since density = mass/volume, a calculator gives us:

$$\text{Density of Liquid} = \frac{\text{Mass of Liquid}}{\text{Volume of Liquid}} = \frac{6.03 \text{ g}}{7.12 \text{ mL}} = 0.8469101 \text{ g/mL}.$$

However, it is not possible for the density to be known to seven significant figures when the measurements from which it is calculated are known to only three significant figures. (Calculators almost always give the wrong number of significant figures -- be aware of this!) At what point should we "round off"?

Since we know that the "3" in "6.03 g" and the "2" in "7.12 mL" are the uncertain digits, let's see what effect a small change in each of these digits has on the calculated density.

$$\frac{6.02 \text{ g}}{7.13 \text{ mL}} = 0.8443197 \text{ g/mL} \qquad \frac{6.04 \text{ g}}{7.11 \text{ mL}} = 0.8495077 \text{ g/mL}$$

Notice that for each of the three calculated densities above, the "0.84" at the beginning remains the same. The first digit which differs (that is, the first uncertain digit) is the digit in the thousandths place -- the third significant digit. (The zero to the left of the decimal point is not significant. Write the number in scientific notation if you're not sure of this!) Therefore, according to the significant figure convention (that is, record all of the certain digits plus one uncertain digit), the "rounding off" should occur after the thousandths place, and the density should be recorded as "0.847 g/mL". Note that three significant digits are present in the measured mass, the measured volume, and the calculated density. This is typical for calculations involving only multiplication and division -- the calculated number has the same number of significant digits as the measured numbers do, if their "sig figs" are the same.

USING SIGNIFICANT FIGURES IN CALCULATIONS (ctd.):

To be sure that a number which has been calculated from measured numbers has the correct number of significant figures, follow the rules below.

A. Multiplication and Division

When multiplying or dividing two numbers, the number of significant figures in the result is the same as the number of significant figures in the measured number which has the _least_ number of significant figures. _Example:_

$$\frac{6.03 \text{ g}}{7.1 \text{ mL}} = 0.8492957 \text{ (calculator)} = 0.85 \text{ (correctly rounded)}$$

Since "6.03" contains _three_ significant digits but "7.1" only has _two_, the result of the division should be a number with only _two_ significant digits.

Problem: What is 6.03 m x 7.1 m ? (_Solution:_ 43 m^2. _Two_ sig figs.)

B. Addition and Subtraction

Since addition and subtraction involve adding or subtracting _columns_ of numbers, the result should include the digit from the _leftmost column which_ _contains an uncertain digit, but no digits after that digit._ _Examples:_

$$
\begin{array}{r}
7.1 \text{ g} \\
+\ 6.03 \text{ g} \\
\hline
13.1 \text{ g}
\end{array}
\qquad
\begin{array}{r}
7.1 \text{ g} \\
-\ 6.03 \text{ g} \\
\hline
1.1 \text{ g}
\end{array}
$$

In each case, the _tenths_ column contains an uncertain digit (the "1" in "7.1"). Hence, no digits past the _tenths_ place are written.

C. Using Defined Numbers

Problem: Express 365.24 feet in yards.

Solution: Since 3 feet = 1 yard _by definition_ (_not_ by measurement!), the "3" in the conversion factor "3 feet/yard" is considered to contain an _infinite_ number of significant figures. (If you like, "3.0000000000...feet/yard!) Therefore, since "365.24 feet" contains _five_ significant figures, the final result should also contain _five_ significant figures. Hence:

$$365.24 \text{ feet} \times \frac{1 \text{ yard}}{3 \text{ feet}} = 121.74666 \text{ (calculator)} = \underline{121.75 \text{ yards}}.$$

ANTOINE LAVOISIER'S COMBUSTION EXPERIMENTS:

The French scientist Antoine Lavoisier has been called the "father of modern chemistry" for his late 18th-century combustion experiments. Combustion is simply the reaction of something with oxygen. This is what happens when something catches fire and burns. Lavoisier was the first person to carefully measure the masses of the materials used and products formed in combustion reactions. Data similar to those obtained by Lavoisier during these experiments are given below.

Expt. No.	Mass of N_2 Used	Mass of O_2 Used	Product and its Mass
1	28.0 g	32.0 g	60.0 g of NO
2	28.0 g	64.0 g	92.0 g of NO_2
3	56.0 g	32.0 g	88.0 g of N_2O

The results of the experiments above and many other experiments like them enabled Lavoisier to determine the following laws of chemical combination.

A. The Law of Conservation of Mass. In a chemical reaction, the total mass present remains constant -- matter is neither created nor destroyed. For example, see Experiment # 1, above. 28.0 g + 32.0 g = 60.0 g.

B. The Law of Definite Proportions. (Also called the Law of Constant Composition.) The relative amount of each element in a particular compound is always the same. For example, the compound NO contains nitrogen and oxygen in a 28:32 mass ratio. If some other ratio is used, the product is not NO!

C. The Law of Multiple Proportions. If two elements can combine so as to form more than one compound, then the masses of one element that react with a fixed mass of the other element will be related by small whole number ratios. For example, compare Experiments # 1 and # 2, above. The mass of N_2 is fixed, and the masses of O_2 which react with it are 32.0 g and 64.0 g -- a ratio of 1:2. Experiments # 1 and # 3 are similar -- a 1:2 ratio of N_2 for a fixed mass of O_2.

DALTON'S ATOMIC THEORY:

In the early part of the 19th century, the English scientist John Dalton proposed a hypothesis to explain Lavoisier's laws of chemical reactions. The main ideas of Dalton's hypothesis are listed below.

a) All elements are composed of small, indivisible, indestructible particles called _atoms_.

b) The atoms of any _one_ element all have the _same_ chemical and physical properties. Atoms of two _different_ elements have _different_ chemical and physical properties.

c) _Molecules_ are formed by joining two or more atoms. Since atoms are indivisible, _whole numbers_ of atoms must be joined in this process -- there's no such thing as a "fraction" of an atom.

d) In a chemical _reaction_, the atoms involved are simply _rearranged_ to form different molecules. _No atoms are created or destroyed_ in a reaction.

This hypothesis was sufficient to explain Lavoisier's laws. Since atoms are neither created nor destroyed during a reaction, no _mass_ can be gained or lost (_Law of Conservation of Mass_). Since atoms are _indivisible_ (that is, they cannot be cut into smaller pieces), they must combine in _whole number ratios_ (_Law of Multiple Proportions_). The Law of Definite Proportions is explained similarly.

Dalton's hypothesis is a shining example of the way science usually works. Dalton's ideas about atoms and molecules are essentially a _model_, whose sole purpose was to explain Lavoisier's results. Dalton's hypothesis has been revised and modified in the last two centuries. Today, for example, we know that atoms _can_ be cut into smaller pieces (_nuclear fission!_), and that not all atoms of an element have identical physical properties (_isotopes!_). But Dalton's main ideas have survived the test of time, and are called the _atomic theory_ today.

THE DISCOVERY OF ELECTRONS:

One of the major revisions that has been made in Dalton's atomic theory is that atoms <u>can</u> be divided into smaller particles. These particles were discovered through a series of experiments done in the late 19th century and early 20th century.

Cathode ray tubes (CRTs) are the "picture tubes" used in television sets and computer terminals. <u>Cathode rays</u> are produced when a high electrical voltage is applied to a glass tube filled with a gas at low pressure. In 1897, the British physicist J. J. Thomson discovered that cathode rays are deflected by external electric and magnetic fields, are attracted to objects which have positive electrical charges, and can cause a small paddlewheel placed in their path to turn. These results allowed Thomson to conclude that cathode rays are composed of small, negatively-charged particles which he called <u>electrons</u>. He reasoned that these electrons were coming from the atoms of the gas in the cathode ray tube. He was also able to determine the ratio of an electron's electrical <u>charge</u> (e) to its <u>mass</u> (m): $e/m = -1.76 \times 10^8$ coulombs/gram. (The <u>coulomb</u> is the SI unit of electrical charge.) Thomson received the 1906 Nobel Prize in Physics.

In 1909, the American physicist Robert Millikan was able to determine the mass of an electron by means of his clever "oil-drop" experiment. Millikan used X-rays to <u>ionize</u> air molecules (that is, to make them give off electrons). The free electrons generated by this process became attached to droplets of oil that Millikan caused to fall between two electrically-charged metal plates. Millikan could make the negatively-charged oil droplets stop falling by adjusting the electrical charges on the metal plates. This enabled him to calculate the electron's electrical <u>charge</u> ($e = -1.60 \times 10^{-19}$ coulombs), and to use Thomson's ratio to calculate the <u>mass</u> of an electron ($m = 9.11 \times 10^{-28}$ grams).

PROTONS AND THE ATOMIC NUCLEUS:

Zinc sulfide glows when charged particles strike it. (This is how the glowing images on television screens are produced!) In 1898, the German physicist Wilhelm Wien found that a sheet of zinc sulfide glows when it is placed behind the negative electrode of a cathode ray tube. From this, Wien concluded that there must be positively-charged particles in cathode ray tubes, since such particles would be attracted to the negatively-charged electrode. Wien was able to measure the masses of these particles, and found that different masses were obtained when different gases were used in the cathode ray tube. The lightest particles came from hydrogen gas, and the particles obtained from other gases had masses which were approximately multiples of the mass of the hydrogen particles. Wien named the hydrogen particles protons, and received the 1911 Nobel Prize in Physics.

In 1911, the British physicist Ernest Rutherford (who had won the 1908 Nobel Prize in Chemistry for his studies of radioactivity) did an experiment in which he directed a beam of alpha particles at a thin sheet of gold foil. (Alpha particles, which are emitted by some radioactive atoms, are positively charged.) Rutherford found that while most of the alpha particles passed directly through the gold foil, a few of them were deflected from the foil at an angle. Some of them were even directed back toward their source, which to Rutherford was like "a cannonball bouncing off of tissue paper"! These results effectively disproved the "plum pudding" model of the atom which had been proposed by J. J. Thomson. This hypothesis stated that electrons were distributed randomly through through a spherical, positively-charged atom. Rutherford's results showed that atoms are mostly empty space, but contain relatively massive positively-charged particles. Rutherford proposed a new model of the atom, in which the positive charges were centered in a massive atomic nucleus, which was surrounded by the electrons.

NEUTRONS, ATOMIC NUMBERS, AND MASS NUMBERS:

After Rutherford proposed his nuclear model of the atom, he did other experiments to try to calculate the _mass_ of the atomic nucleus. He tried to do this by measuring the _charge_ of the atomic nucleus and calculating the mass based on Wien's values for the mass and charge of the proton, but these calculations gave results that were approximately _half_ of the masses of the atomic nuclei that he was able to measure experimentally. Rutherford concluded that atomic nuclei contain not only protons, but also particles with about the same masses as protons but with no electrical charge. These particles were called _neutrons_, because they were neither positively-charged nor negatively-charged -- they were _neutral_. (In 1935, the British physicist James Chadwick won the Nobel Prize in Physics for his experiments which showed clearly that atomic nuclei contain neutrons.)

The table below summarizes the results of Thomson's, Millikan's, Wien's, and Rutherford's experiments. Masses are given in _atomic mass units (amu)_ and electrical charges are based on a value of "+1.0" for the proton. The actual charge on a proton is $+1.6 \times 10^{-19}$ coulombs, and 1.0 amu = 1.66×10^{-24} grams.

Subatomic Particle	Mass, amu	Charge	Location in Atom
Proton	1.0	+1.0	center (nucleus)
Neutron	1.0	0.0	center (nucleus)
Electron	"0.0"	-1.0	outside the nucleus

Many physical and chemical properties of atoms are determined by the _number of electrons_ surrounding the nucleus. This number will be the same as the _number of protons_ in the nucleus if the atom is electrically neutral (as most atoms are). This number is called the _atomic number_ of the atom. The _mass number_ of an atom is simply the number of "massive" particles (protons and neutrons) its nucleus contains.

ISOTOPES, ATOMIC WEIGHTS, AND MOLECULAR WEIGHTS:

A major modification of Dalton's atomic theory involves the existence of isotopes. Isotopes are atoms of the same element (and therefore, atoms which have the same atomic number) which have different physical properties. Specifically, two isotopes differ in their masses by virtue of the fact that they contain different numbers of neutrons. Isotopes are usually represented by symbols of the general form $^A_Z Q$, where Q is the chemical symbol for the element, Z is its atomic number, and A is its mass number. For example, an atom of carbon-14 ($^{14}_6 C$) contains six protons, six electrons, and eight (14 - 6) neutrons. Another isotope of carbon, carbon-12 ($^{12}_6 C$), contains only six neutrons per atom.

The atomic weight of an element is the average mass of its atoms. We can speak of average masses because different isotopes occur naturally to various extents. For example, naturally-occurring chlorine is about 75% chlorine-35 ($^{35}_{17} Cl$) and 25% chlorine-37 ($^{37}_{17} Cl$). The atomic weight of chlorine is approximately 35.5 amu, since (0.75)(35 amu) + (0.25)(37 amu) = 35.5 amu. The atomic weight and atomic number of each element can be found near its symbol on the Periodic Table.

An important property of a compound is its molecular weight (sometimes called its formula weight). The molecular weight of a compound is simply the sum of the atomic weights of the atoms in its formula. As an example, the molecular weight of sodium chloride (NaCl, table salt) is simply the sum of the atomic weights of sodium (Na = 22.98977 amu) and chlorine (Cl = 35.453 amu) -- that is, 58.443 amu. The molecular weight of water (H_2O) is found by adding the atomic weight of oxygen (O = 15.9994 amu) to twice the atomic weight of hydrogen (2 x H = 2 x 1.0079 amu = 2.0158 amu) -- the result is 18.0152 amu. The molecular weight of sucrose (table sugar, $C_{12}H_{22}O_{11}$) is calculated similarly (carbon = 12.011 amu): (12 x 12.011 amu) + (22 x 1.0079 amu) + (11 x 15.9994 amu) = 342.299 amu.

                              THE MOLE CONCEPT:

How many socks are there in a <u>pair</u> of socks? Right -- <u>two</u>.

Slightly harder, now: How many 3¢-stamps are there in a <u>dozen</u>?
Right -- <u>twelve</u>! (The same number of things there are in a dozen of <u>anything</u>!)

"Pair" and "dozen" are just <u>words that represent numbers</u>. "Pair" represents the number <u>two</u>; "dozen" represents the number <u>twelve</u>. Chemists use the word <u>"mole"</u> to represent a very large number -- 6.02×10^{23}. (This number is sometimes called <u>Avogadro's Number</u>, after the Italian scientist Amadeo Avogadro.) This number is so huge that it's useless for counting large things like socks or stamps. What it <u>is</u> useful for is counting small things like <u>atoms</u> and <u>molecules</u>.

The mole concept allows us to think about substances on both the large (visible) scale and the small (sub-microscopic) scale simultaneously. This is due to the fact that <u>one mole of atoms</u> of an element has the same <u>mass</u> as the <u>atomic weight</u> of the element, expressed in <u>grams</u>. For example, the element <u>neon (Ne)</u> has an atomic weight of 20.179 amu. Another way of saying this is to state that a sample of neon weighing <u>20.179 grams</u> contains <u>one mole (6.02×10^{23})</u> of Ne atoms.

Since molecular weights are simply the sums of atomic weights, it is also true that <u>one mole of molecules</u> of a compound has the same <u>mass</u> as the <u>molecular weight</u> of the compound, expressed in <u>grams</u>. For example, <u>water (H_2O)</u> has a molecular weight of 18.0152 amu. Therefore, <u>one mole of water molecules</u> has a mass of <u>18.0152 grams</u>. <u>Sucrose</u> (table sugar, $C_{12}H_{22}O_{11}$) has a molecular weight of 342.299 amu, so <u>Avogadro's Number of sucrose molecules weighs 342.299 grams</u>. The molecular weight of <u>sodium chloride</u> (table salt, NaCl) is 58.443 amu; sodium chloride contains <u>58.443 grams per mole</u>. (When referring to atomic weights or molecular weights, the units "amu" or "grams per mole" (sometimes written "g/mol") may be used interchangeably.)

USING THE MOLE CONCEPT IN CHEMICAL CALCULATIONS:

Problem: What <u>mass</u> of argon gas equals 0.375 moles of argon atoms?

Solution: Argon (Ar) has an atomic weight of 39.948 amu. Therefore, the conversion equation is: 39.948 grams of Ar = 1 mole of Ar atoms. Setting up the appropriate conversion factor, we obtain the following result:

$$0.375 \text{ moles Ar} \times \frac{39.948 \text{ g Ar}}{1 \text{ mole Ar}} = 15.0 \text{ g Ar.}$$

Problem: How many <u>atoms</u> of lead are present in 50.0 grams of lead?

Solution: Lead (Pb) has an atomic weight of 207.2 amu (207.2 <u>g/mole</u>). The first step is to use this conversion factor to obtain an amount in <u>moles</u>:

$$50.0 \text{ g Pb} \times \frac{1 \text{ mol Pb}}{207.2 \text{ g Pb}} = 0.241 \text{ moles Pb.}$$

Now, use Avogadro's Number to convert this amount into a number of atoms. The conversion equation is: 1 mole = 6.02×10^{23} atoms. Therefore:

$$0.241 \text{ moles Pb} \times \frac{6.02 \times 10^{23} \text{ Pb atoms}}{1 \text{ mole Pb}} = 1.45 \times 10^{23} \text{ Pb atoms.}$$

Problem: A sample of <u>methane</u> (CH_4) contains 3.75×10^{24} hydrogen atoms. Calculate the <u>mass</u> of the sample of methane.

Solution: Since each methane <u>molecule</u> contains <u>four</u> hydrogen atoms, we can find the number of methane molecules using the following conversion equation: 1 CH_4 molecule = 4 H atoms. Using this as a conversion factor, we get:

$$3.75 \times 10^{24} \text{ H atoms} \times \frac{1 \text{ } CH_4 \text{ molecule}}{4 \text{ H atoms}} = 9.38 \times 10^{23} \text{ } CH_4 \text{ molecules.}$$

Now, use Avogadro's Number to convert this to an amount in <u>moles</u>:

$$9.38 \times 10^{23} \text{ } CH_4 \text{ molecules} \times \frac{1 \text{ mole } CH_4}{6.02 \times 10^{23} \text{ molecules}} = 1.56 \text{ moles } CH_4.$$

Finally, use methane's <u>molecular weight</u> to convert this to a <u>mass</u>:

$$1.56 \text{ moles } CH_4 \times \frac{16.043 \text{ g } CH_4}{1 \text{ mole } CH_4} = 25.0 \text{ grams of } CH_4.$$

19

DETERMINATION OF A FORMULA FROM PERCENTAGE COMPOSITION DATA:

The <u>percentage by mass</u> of an element in a compound is the same as the mass (in grams) of that element needed to form 100.00 grams of the compound, except it's expressed as a percentage. As an example, consider sodium chloride: one mole (58.443 g) of NaCl is made from one mole of Na (22.98977 g) and one mole of Cl (35.453 g), so the percentages of Na and Cl in NaCl are easily calculated:

$$\% \text{ Na} = \frac{22.98977 \text{ g Na}}{58.443 \text{ g NaCl}} \times 100.00\% = 39.337\% \text{ Na by mass.}$$

$$\% \text{ Cl} = \frac{35.453 \text{ g Cl}}{58.443 \text{ g NaCl}} \times 100.00\% = 60.663\% \text{ Cl by mass.}$$

Therefore, to prepare 100.000 grams of sodium chloride, 39.337 grams of sodium and 60.663 grams of chlorine are needed. (Note that <u>mass is conserved!</u>)

<u>Problem</u>: Calculate the percentages by mass of carbon and hydrogen in the compound <u>methane</u> (CH_4). (Approximate Solution: 75% C, 25% H.)

Usually, the percentages by mass of the elements in a compound are determined <u>experimentally</u>, and then the <u>formula</u> of the compound is calculated from the percentage composition data.

<u>Problem</u>: The compound <u>isopropyl alcohol</u> ("rubbing alcohol") contains 60.0% C, 13.4% H, and 26.6% O by mass. What is the formula of isopropyl alcohol?

<u>Solution</u>: The percentages tell us that 100.0 g of alcohol is made from 60.0 g of C, 13.4 g of H, and 26.6 g of O. Convert these to <u>moles</u> of atoms:

$$\frac{60.0 \text{ g C atoms}}{12.011 \text{ g/mole C}} = 5.00 \text{ moles C} \qquad \frac{13.4 \text{ g H atoms}}{1.0079 \text{ g/mole H}} = 13.3 \text{ moles H}$$

$$\frac{26.6 \text{ g O atoms}}{15.9994 \text{ g/mole O}} = 1.66 \text{ moles O} \qquad \text{Formula} = C_{5.00}H_{13.3}O_{1.66} \quad (???)$$

Dalton's atomic theory says that we can't have .3 or .66 of an atom, so we need <u>whole numbers</u> with a 5.00/13.3/1.66 ratio. Dividing each subscript by the <u>lowest</u> subscript present (here, 1.66) gives:

$$C_{\frac{5.00}{1.66}}H_{\frac{13.3}{1.66}}O_{\frac{1.66}{1.66}} = C_{3.01}H_{8.01}O_{1.00} = \underline{C_3H_8O.}$$

EMPIRICAL FORMULAS AND MOLECULAR FORMULAS:

Percentage composition data can only provide the empirical formula of a compound -- that is, the formula which expresses the $\underline{ratios}$ of the atoms present in the compound using the $\underline{smallest\ whole\ numbers\ possible}$. To completely determine the $\underline{molecular\ formula}$ of a compound, the $\underline{molecular\ weight}$ must be determined and compared to the "empirical formula weight". Examples appear below:

Name of Compound	Empirical Formula	Empirical Formula Weight	Molecular Weight	Molecular Formula
Isopropyl Alcohol	C_3H_8O	60 g/mole	60 g/mole	C_3H_8O
Acetylene	CH	13 g/mole	26 g/mole	C_2H_2
Glucose	CH_2O	30 g/mole	180 g/mole	$C_6H_{12}O_6$

For isopropyl alcohol, the empirical formula weight is the $\underline{same}$ as the molecular weight, so the empirical formula is the same as the molecular formula. For acetylene, the two formula weights differ by a factor of $\underline{two}$, so the molecular formula is just $\underline{twice}$ the empirical formula -- $(CH)_2 = C_2H_2$. For glucose, 180/30 = $\underline{6}$, so the molecular formula is $(CH_2O)_6 = C_6H_{12}O_6$.

$\underline{Problem}$: $\underline{Butane}$ (the fluid in cigarette lighters), contains 82.66% C and 17.34% H by mass. The molecular weight of butane is approximately 58 g/mole. Calculate the $\underline{molecular\ formula}$ of butane.

$\underline{Solution}$: 100.00 g of butane = 82.66 g of C and 17.34 g of H. Moles: 82.66/12.011 = 6.882 moles C, 17.34/1.0079 = 17.20 moles H. $C_{6.882}H_{17.20}$ can't be right, so dividing by 6.882 gives $C_{1.000}H_{2.500}$. Multiplying by $\underline{2}$ gives us whole numbers: C_2H_5 = empirical formula = 29 g/mole. Molecular formula = C_4H_{10}.

$\underline{Alternative\ Solution}$: Multiply 58 g of butane by the percentages: 58 g x .8266 = 48 g of C, 58 g x .1734 = 10 g of H. Convert these into $\underline{moles}$: $\frac{48\ g\ of\ C}{12\ g/mole}$ = 4 moles of C, $\frac{10\ g\ of\ H}{1\ g/mole}$ = 10 moles of H. Molecular formula = C_4H_{10}.

MOLECULES, IONS, AND POLYATOMIC IONS:

Some compounds are made of <u>molecules</u>, which are electrically <u>neutral</u> collections of two or more atoms that are joined in some way. These compounds are called <u>covalent compounds</u>, and are represented by <u>molecular formulas</u>. Water (H_2O) and sucrose ($C_{12}H_{22}O_{11}$, table sugar) are examples of covalent compounds. Some <u>elements</u> are also molecular in nature. For example, the <u>oxygen</u> in the air we breathe occurs naturally as molecules with the molecular formula O_2.

Other compounds are made of <u>ions</u>, which are electrically <u>charged</u> particles made from one or more atoms. These compounds are called <u>ionic compounds</u>, and are represented by <u>empirical formulas</u>. (There's no such thing as a "molecule" of an ionic compound, so it doesn't make much sense to refer to its formula as a "molecular" formula.) An example of an ionic compound is sodium chloride (NaCl, table salt), which is made up of two kinds of ions: Na^+ and Cl^-. Notice that the <u>charge</u> of the ion is written as a superscript. Ions with a positive charge are called <u>cations</u>; ions with a negative charge are called <u>anions</u>.

<u>Polyatomic ions</u> are ions which are made from more than one atom. Examples include the <u>hydroxide ion</u>, OH^-, which is found in <u>lye</u> (sodium hydroxide, NaOH), and the <u>carbonate ion</u>, CO_3^{2-}, which is found in <u>limestone</u> (calcium carbonate, $CaCO_3$). Some polyatomic ions contain <u>three</u> or more elements. An example is the <u>acetate ion</u>, $C_2H_3O_2^-$, which is found in <u>vinegar</u> (acetic acid, $HC_2H_3O_2$). Polyatomic ions may be either <u>anions</u> (like those above) or <u>cations</u>. An example of a polyatomic cation is the <u>ammonium ion</u>, NH_4^+, which is found in the fertilizer ammonium nitrate (NH_4NO_3, which also contains the <u>nitrate ion</u>, NO_3^-). Other common polyatomic ions include the <u>sulfate ion</u> (SO_4^{2-}), the <u>phosphate ion</u> (PO_4^{3-}), and the <u>bicarbonate ion</u> (HCO_3^-, which is found in baking soda (sodium bicarbonate, $NaHCO_3$)).

WRITING FORMULAS FOR IONIC COMPOUNDS:

The formulas of ionic compounds are written as _empirical formulas_ --
that is, the smallest whole numbers are used which accurately describe the ratios
of the ions present. For example, the formula for sodium chloride is written as
$NaCl$, not Na_2Cl_2 or Na_3Cl_3. Simply "NaCl" is enough to show the one-to-one ratio
of the sodium ions (Na^+) to the chloride ions (Cl^-).

Two things should be noted here. First, when chemists write the
formulas of ionic compounds, they almost always write the symbol for the _cation_
first and the symbol for the _anion last_. (Sodium chloride is "NaCl", not "ClNa".
"ClNa" is not "wrong", but chemists prefer to write the symbol for the positive
ion first to avoid confusion about which ion has which charge.) Second, the net
electrical charge implied by the compound's formula is _zero_ -- that is, the total
of the positive charges is exactly cancelled by the total of the negative charges.
This _must_ be true for any ionic compound, since all _compounds_ are electrically
neutral. (In sodium chloride, the +1 charge of Na^+ is exactly cancelled by the
-1 charge of Cl^-. This explains why the formula of sodium chloride is NaCl, not
$NaCl_2$ or Na_2Cl. On the other hand, _calcium chloride_ has the formula $CaCl_2$ due to
the fact that calcium ions have a charge of +2. Therefore, _two_ Cl^- ions are
needed to exactly cancel _one_ Ca^{2+} ion, and the formula is $CaCl_2$.)

When writing the formulas of compounds which contain _polyatomic ions_,
it's a good idea to put _parentheses_ around the formula of the polyatomic ion,
especially if _more than one_ polyatomic ion is present in the formula. For
example, calcium carbonate is written as $CaCO_3$ (since only _one_ CO_3^{2-} ion is
present), but calcium hydroxide is written as $Ca(OH)_2$ (since _two_ OH^- ions are
present). Writing calcium hydroxide as CaO_2H_2 is not "wrong", but it's not as
descriptive as $Ca(OH)_2$ -- the parentheses clearly show the presence of OH^- ions.

ACIDS, BASES, AND CHEMICAL REACTIONS:

Compounds which generate hydrogen ions (H^+) are called <u>acids</u>. Some well-known acids include <u>hydrochloric acid</u> (HCl, "stomach acid"), <u>sulfuric acid</u> (H_2SO_4, found in "acid rain"), and <u>acetic acid</u> ($HC_2H_3O_2$, vinegar). Notice that the formulas of these acids all begin with "H". This is usually a good "tip-off" that a compound might be an acid.

Compounds which generate hydroxide ions (OH^-) are called <u>bases</u> or <u>alkalis</u>. Some well-known bases include sodium hydroxide (NaOH, "lye") and magnesium hydroxide ($Mg(OH)_2$, "milk of magnesia"). Notice that the formulas of these bases both end with "OH". For <u>ionic</u> compounds, this is a good "tip-off" that a base is present.

Acids and bases tend to <u>neutralize</u> each other when they are mixed together -- that is, they tend to form new substances which are neither acidic nor basic. This is one example of a <u>chemical reaction</u> -- the changing of one or more substances into one or more new substances. Typically, an acid-base reaction consists of the H^+ ions from the acid neutralizing the OH^- ions from the base by combining with them to form molecules of <u>water</u> (HOH, or H_2O). For example, the reaction of hydrochloric acid with sodium hydroxide forms water and sodium chloride -- table salt. This reaction is summarized in the <u>chemical equation</u>:

$$HCl \ + \ NaOH \ \longrightarrow \ H_2O \ + \ NaCl$$

In this reaction, HCl and NaOH are referred to as the <u>reactants</u> -- the substances which react with each other. The <u>products</u> (the new substances which are formed) are H_2O and NaCl. The arrow indicates the progress of the reaction, and can be read as "forms" or "produces" or "yields" or "gives". One way to "read" the above equation is to say that "hydrochloric acid and sodium hydroxide produce water and sodium chloride."

WRITING BALANCED EQUATIONS FOR CHEMICAL REACTIONS:

Suppose we wanted to write a chemical equation which describes the reaction between hydrogen gas and oxygen gas to form liquid water. One way to do this might be to simply write: $H + O \longrightarrow H_2O$. However, this isn't the most best way to describe this reaction, since hydrogen and oxygen normally occur as <u>diatomic molecules</u> -- H_2 and O_2. Therefore, a somewhat better way to write this equation would be: $H_2 + O_2 \longrightarrow H_2O$. But even this equation isn't perfect. Notice that there are <u>two</u> oxygen atoms among the <u>reactants</u>, but only <u>one</u> oxygen atom among the <u>products</u>. This seems to imply that an oxygen atom has "vanished into thin air", which is impossible according to Dalton's atomic theory.

The best way to write the above reaction is: $2\ H_2 + O_2 \longrightarrow 2\ H_2O$. This equation states that <u>two</u> molecules of H_2 react with <u>one</u> molecule of O_2, producing <u>two</u> molecules of H_2O. (The "2" in front of the H_2 and the H_2O is called the <u>coefficient</u> of that formula. When the coefficient is "1", as it is for O_2 in this case, the "1" is usually omitted.) Note that in this equation, there are <u>four</u> hydrogen atoms among the reactants <u>and</u> the products, and <u>two</u> oxygen atoms are also found on <u>both</u> sides of the arrow. When the numbers of atoms of each element found on both sides of the arrow are the <u>same</u>, the equation is called <u>"balanced"</u>.

The usual way to balance an equation is to write the formulas first, then change the coefficients. (In the above case, we could have "balanced" the equation by changing a formula -- $H_2 + O_2 \longrightarrow H_2O_2$ -- but this is the equation of a <u>different reaction</u>, not the formation of <u>water</u>!)

<u>Problem</u>: Balance the following equation: $Fe + Cl_2 \longrightarrow FeCl_3$.

<u>Solution</u>: The reactants contain one less chlorine atom than the products do. Placing a "3" in front of the Cl_2 and a "2" in front of the $FeCl_3$ corrects this. Placing a "2" in front of the Fe gives: $2\ Fe + 3\ Cl_2 \longrightarrow 2\ FeCl_3$.

QUALITATIVE ANALYSIS AND QUANTITATIVE ANALYSIS:

The analysis of a new or unknown substance is often done in two steps: qualitative analysis and quantitative analysis. Qualitative analysis is usually done first, since it is the process of discovering whether or not a particular substance (element, compound, etc.) is present in the sample being analyzed. The answer to the questions posed by qualitative analysis is always either yes or no -- a particular substance is either present or absent in the sample being studied.

Quantitative analysis is more mathematical, because it is the process of discovering the relative amounts of the various substances present in the sample being studied. The answers to the questions posed by quantitative analysis are always numbers. (Note: "Amounts" of pure substances are measured in moles.)

Organic chemists discover many new organic compounds every year. Some of these compounds prove effective in treating many of the diseases that plague humanity. Many of these new compounds are composed only of carbon, hydrogen, and oxygen. A particularly useful quantitative analysis experiment for this kind of compound is combustion analysis. In this process, the unknown compound is placed in an oven and heated to high temperatures in the presence of excess oxygen gas. Combustion (burning) occurs, completely converting the organic compound into water and carbon dioxide. The equation for this reaction is given below.

$$C_xH_yO_z \; + \; \text{excess } O_2 \; \longrightarrow \; x \; CO_2 \; + \; \frac{y}{2} \; H_2O$$

The carbon dioxide and water vapor produced in this process are passed through the combustion apparatus and collected in separate chambers, which are then weighed. From the masses of carbon dioxide and water formed in this process, the amounts of carbon and hydrogen present in the sample can be determined. (The combustion analysis experiment is sometimes called "C and H analysis".) This can lead to the determination of the molecular formula of the unknown compound.

SOLVING PROBLEMS IN COMBUSTION ANALYSIS:

Problem: Ascorbic acid (Vitamin C) has a molecular weight of 176 g/mole. A sample of ascorbic acid weighing 5.11 mg gave 7.66 mg of CO_2 and 2.09 mg of H_2O upon combustion. What is the molecular formula of ascorbic acid?

Solution: Step 1 -- Convert the masses of CO_2 and H_2O into amounts in moles. (Note: 1 mg = 1 x 10^{-3} g, so 7.66 mg = 7.66 x 10^{-3} g, etc.)

$$\frac{7.66 \times 10^{-3} \text{ g } CO_2}{44.01 \text{ g/mole } CO_2} = 1.74 \times 10^{-4} \text{ moles } CO_2 = 1.74 \times 10^{-4} \text{ moles C atoms.}$$

$$\frac{2.09 \times 10^{-3} \text{ g } H_2O}{18.02 \text{ g/mole } H_2O} = 1.16 \times 10^{-4} \text{ moles } H_2O = 2.32 \times 10^{-4} \text{ moles H atoms.}$$

(Note: The amount of H_2O present is multiplied by two, since each molecule of water contains two hydrogen atoms.)

Step 2 -- Convert the amounts of C and H back to masses and subtract them from the mass of the sample to determine the mass of oxygen in the sample.

1.74 x 10^{-4} moles C x 12.011 g/mole = 2.09 x 10^{-3} g C in sample.

2.32 x 10^{-4} moles H x 1.0079 g/mole = 2.34 x 10^{-4} g H in sample.

5.11 x 10^{-3} g sample - 2.09 x 10^{-3} g C - 2.34 x 10^{-4} g H =

$$\frac{2.79 \times 10^{-3} \text{ g O}}{15.9994 \text{ g/mol O}} = 1.74 \times 10^{-4} \text{ moles O atoms.} \qquad 2.79 \times 10^{-3} \text{ g O.}$$

Step 3 -- Solve for the empirical formula of the compound.

$$C_{\frac{0.000174}{0.000174}} H_{\frac{0.000232}{0.000174}} O_{\frac{0.000174}{0.000174}} = C_{1.00} H_{1.33} O_{1.00} = C_{3.00} H_{3.99} O_{3.00}$$
$$= C_3 H_4 O_3.$$

(Multiplying by three gives whole numbers.)

Step 4 -- Use the molecular weight to determine the molecular formula.

$C_3H_4O_3$ = (3 x 12.011) + (4 x 1.0079) + (3 x 15.9994) = 88 g/mole.

$\frac{176 \text{ g/mole}}{88 \text{ g/mole}}$ = 2, so the molecular formula = $(C_3H_4O_3)_2 = C_6H_8O_6$.

AN ALTERNATIVE METHOD FOR SOLVING PROBLEMS IN COMBUSTION ANALYSIS:

Problem: Ascorbic acid (Vitamin C) has a molecular weight of 176 g/mole. A sample of ascorbic acid weighing 5.11 mg gave 7.66 mg of CO_2 and 2.09 mg of H_2O upon combustion. What is the molecular formula of ascorbic acid?

Solution: Step 1 -- Use the Law of Conservation of Mass to find the mass of oxygen gas (O_2) consumed in the reaction.

$$\text{mass of } O_2 = \text{mass of } CO_2 + \text{mass of } H_2O - \text{mass of sample}$$

$$= 7.66 \text{ mg} + 2.09 \text{ mg} - 5.11 \text{ mg} = \underline{4.64 \text{ mg of } O_2 \text{ consumed.}}$$

Step 2 -- Convert all masses into moles. (Note: $1 \text{ mg} = 1 \times 10^{-3}$ g.)

$$\frac{5.11 \times 10^{-3} \text{ g sample}}{176 \text{ g/mole sample}} = 2.90 \times 10^{-5} \text{ moles sample.}$$

$$\frac{4.64 \times 10^{-3} \text{ g } O_2}{32.00 \text{ g/mole } O_2} = 1.45 \times 10^{-4} \text{ moles } O_2.$$

$$\frac{7.66 \times 10^{-3} \text{ g } CO_2}{44.01 \text{ g/mole } CO_2} = 1.74 \times 10^{-4} \text{ moles } CO_2.$$

$$\frac{2.09 \times 10^{-3} \text{ g } H_2O}{18.02 \text{ g/mole } H_2O} = 1.16 \times 10^{-4} \text{ moles } H_2O.$$

Step 3 -- Divide each amount above by the lowest amount present.

Sample: $\dfrac{0.0000290}{0.0000290} = 1.00.$ CO_2: $\dfrac{0.000174}{0.0000290} = 6.00.$

O_2: $\dfrac{0.000145}{0.0000290} = 5.00.$ H_2O: $\dfrac{0.000116}{0.0000290} = 4.00.$

Step 4 -- Write the equation which describes the combustion reaction, using the numbers calculated in Step 3 as the coefficients.

$$C_x H_y O_z + 5 O_2 \longrightarrow 6 CO_2 + 4 H_2O$$

Step 5 -- Assign numbers to x, y, and z to give a balanced equation.

$x = 6.$ $y = 4 \times 2 = 8.$ Molecular Formula = $\underline{C_6 H_8 O_6}.$

$z = (6 \times 2) + (4 \times 1) - (5 \times 2) = 6.$

COOKING, CHEMISTRY, AND STOICHIOMETRY:

Consider the following shortbread recipe[1]:

"In a bowl, cream 1 cup of softened butter and
1/2 cup of sifted powdered sugar until light and fluffy.
Stir in 2 cups of all-purpose flour and blend well.
Gather up the dough and press it evenly over the bottom
of a lightly-greased 9-inch square baking pan. Bake for
45 minutes at 325 °F. Transfer pan to wire rack and cut
into 25 squares. Let cool in pan."

Since the above process describes chemical reactions -- after all, substances _are_ being converted into new substances -- a chemist might record the information above using a chemical equation such as the one given below:

$$1 \text{ Bu} + 1/2 \text{ Su} + 2 \text{ Fl} \xrightarrow[45 \text{ min}]{163 \text{ °C}} 25 \text{ Sq}$$

where "Bu" = butter, "Su" = sugar, "Fl" = flour, and "Sq" = shortbread squares!

The equation above gives the same information as the recipe does about the _reactions_ involved -- the relative amounts of the ingredients to use, the "reaction conditions" (temperature and baking time), and the number of shortbread squares produced -- but the recipe is obviously more useful in the kitchen!

The _stoichiometry_ of a chemical reaction is simply the quantitative description of the relative amounts of the substances involved in the reaction. The difference for chemists is that "amounts" are measured in _moles_, not cups! Consider the following balanced equation: $PCl_5 + 4 H_2O \longrightarrow H_3PO_4 + 5 HCl$. This equation tells us that one PCl_5 molecule and four water molecules will react to give one H_3PO_4 molecule and five HCl molecules. It also tells us that one _mole_ of PCl_5 and four _moles_ of water react to give one _mole_ of H_3PO_4 and five _moles_ of HCl. The balanced equation therefore allows us to think on both the small scale (molecules) and the large scale (moles) simultaneously. Of course, to actually _do_ this reaction, a recipe (or "experimental procedure") would come in handy in lab!

---

1) "Easy Basics for Good Cooking", Sunset Books, Lane Publishing Co., Menlo Park, CA   94025.

SOLVING STOICHIOMETRY PROBLEMS:

The problem with chemical reactions is that they take place on the molecular level -- that is, only whole numbers of molecules react. Balanced chemical equations give us information about the numbers of molecules that react and the numbers of molecules that are formed -- that is, about the stoichiometry of the reaction. The problem is that in the laboratory, we can't measure out a particular number of molecules directly. All we can do is to measure out masses (in grams) of solids or liquids, or volumes (in liters) of liquids. Hence, we need to be able to convert these units into amounts in moles of molecules (and back again!) in order to be able to use the data that we obtain in the lab.

Typically, problems involving the stoichiometry of reactions are solved in the way outlined below. Consider the generic balanced equation below:

$$w\ A\ +\ x\ B\ \longrightarrow\ y\ C\ +\ z\ D$$

(This is to be read as "w moles of A and x moles of B react to form y moles of C and z moles of D". The equation must be balanced, since only the balanced equation represents the true ratios in which molecules react and form.)

Suppose you weigh out some mass of compound A and you want to find out what mass of compound B you need to weigh out to react completely with your sample of A. The first thing to do is to convert your mass of A (in grams) into a number of moles of A. This should be easy -- just divide by the molecular weight of A, in g/mole! Then, since A and B react in a mole ratio of w:x (look at the balanced equation!), multiply the amount of A (in moles) by x/w to obtain the amount of B (in moles) needed to react completely with your sample of A. (Here, the balanced equation is the conversion equation, and x/w is the conversion factor!) The final step is to convert this quantity back to a mass (in grams), which can be done by simply multiplying by the molecular weight of B (in g/mole).

$$\text{A TYPICAL STOICHIOMETRY PROBLEM:}$$

Problem: What mass of O_2 is needed to react completely with 56.8 g of NH_3 ? What masses of NO and H_2O are formed? The balanced equation is:

$$4\ NH_3\ +\ 5\ O_2\ \longrightarrow\ 4\ NO\ +\ 6\ H_2O$$

Solution: To find the mass of O_2 needed, first convert NH_3 to moles:

$$56.8\ g\ NH_3 \times \frac{1\ mole\ NH_3}{17.0304\ g\ NH_3} = 3.34\ moles\ NH_3.$$

Next, use the balanced equation to find the amount of O_2 needed:

$$3.34\ moles\ NH_3 \times \frac{5\ moles\ O_2}{4\ moles\ NH_3} = 4.17\ moles\ O_2.$$

Finally, convert this amount to a mass:

$$4.17\ moles\ O_2 \times \frac{31.9988\ g\ O_2}{1\ mole\ O_2} = 133\ g\ O_2. \quad (\underline{Note}:\ 3\ \text{"sig figs"}!)$$

The mass of NO formed is found in a similar fashion. The first step is the same as the first step above -- 3.34 moles of NH_3 are reacted. From there:

$$3.34\ moles\ NH_3 \times \frac{4\ moles\ NO}{4\ moles\ NH_3} = 3.34\ moles\ NO.$$

$$3.34\ moles\ NO \times \frac{30.0061\ g\ NO}{1\ mole\ NO} = 100\ g\ NO. \quad (Again,\ 3\ \text{"sig figs"}!)$$

The mass of H_2O formed is found similarly. This time, only the conversion factors are shown, but the stepwise logic is the same!

$$56.8\ g\ NH_3 \times \frac{1\ mole\ NH_3}{17.0304\ g\ NH_3} \times \frac{6\ moles\ H_2O}{4\ moles\ NH_3} \times \frac{18.0152\ g\ H_2O}{1\ mole\ H_2O} = 90.1\ g\ H_2O.$$

To check your work, apply the <u>Law of Conservation of Mass</u>. The total mass of the products should be the same as the total mass of the reactants!

$$\begin{array}{lll}
56.8\ g\ NH_3 & 100\quad g\ NO & \text{(The masses are equal, given that}\\
\underline{+\ 133\quad g\ O_2} & \underline{+\ 90.1\ g\ H_2O} & \text{each mass only contains } \underline{three}\\
190\quad g\ reactants & 190\ g\ products & \underline{\text{significant figures}}!)
\end{array}$$

$$\text{LIMITING REACTANTS:}$$

Problem: How much $FeCl_3$ can be formed from the reaction of 10.0 g of Fe and 20.0 g of Cl_2 ? The balanced equation is: $2\ Fe\ +\ 3\ Cl_2 \longrightarrow 2\ FeCl_3$.

Solution: <u>29.0 grams!</u> <u>Not</u> 30.0 grams (10.0 g + 20.0 g). <u>Why?</u>

Consider this as a stoichiometry problem: What mass of Cl_2 is needed to react with 10.0 g of iron? The calculations are shown below.

$$10.0 \text{ g Fe} \times \frac{1 \text{ mole Fe}}{55.847 \text{ g Fe}} \times \frac{3 \text{ moles } Cl_2}{2 \text{ moles Fe}} \times \frac{70.906 \text{ g } Cl_2}{1 \text{ mole } Cl_2} = 19.0 \text{ g } Cl_2.$$

What happens in the reaction is that the 10.0 g of iron reacts with the 19.0 g of chlorine calculated above, giving <u>29.0 g</u> of $FeCl_3$ and leaving <u>1.0 g</u> of chlorine <u>unreacted</u>, since there is no more iron for it to react with. In this case, iron is referred to as the <u>limiting reactant</u> (or <u>limiting reagent</u>), since it is used up before all of the chlorine can react and thereby sets an <u>upper limit</u> on the amount of $FeCl_3$ that can be formed.

Problem: For the reaction above, if 10.0 grams of iron and 10.0 grams of chlorine are allowed to react, how much $FeCl_3$ can be formed?

Solution: The above calculation shows that 10.0 g of Fe requires 19.0 g of Cl_2 to react completely. Since only 10.0 g of Cl_2 are available here, the Cl_2 will be consumed <u>first</u>. Therefore, Cl_2 is the limiting reagent. The $FeCl_3$ calculation <u>must</u> be based on the amount of the <u>limiting</u> reagent present. Hence:

$$10.0 \text{ g } Cl_2 \times \frac{1 \text{ mole } Cl_2}{70.906 \text{ g } Cl_2} \times \frac{2 \text{ moles } FeCl_3}{3 \text{ moles } Cl_2} \times \frac{162.206 \text{ g } FeCl_3}{1 \text{ mole } FeCl_3} = 15.3 \text{ g } FeCl_3.$$

Problem: For the reaction above, if 10.0 g of iron and 15.0 g of chlorine are allowed to react, which is the limiting reagent?

Solution: Don't be fooled by the masses! 15.0 g is <u>less</u> than the 19.0 g calculated above, so Cl_2 is the limiting reagent in this case.

$$\text{THEORETICAL YIELD AND PERCENTAGE YIELD:}$$

The <u>theoretical yield</u> of a reaction refers to the amount of product that can be formed if the <u>limiting reactant</u> is <u>completely converted</u> into the product. Theoretical yields are expressed in either <u>grams</u> or <u>moles</u>.

<u>Problem</u>: Calculate the theoretical yield of $FeCl_3$ if 10.0 grams of Fe and 15.0 grams of Cl_2 are allowed to react. The balanced equation is:

$$2\ Fe\ +\ 3\ Cl_2\ \longrightarrow\ 2\ FeCl_3$$

<u>Solution</u>: Which is the limiting reactant? In this case, it's Cl_2:

$$15.0\ g\ Cl_2\ \times\ \frac{1\ mole\ Cl_2}{70.906\ g\ Cl_2}\ \times\ \frac{2\ moles\ Fe}{3\ moles\ Cl_2}\ \times\ \frac{55.847\ g\ Fe}{1\ mole\ Fe}\ =\ 7.88\ g\ Fe.$$

This is the amount of iron needed to react with 15.0 g of chlorine. 10.0 g of iron is available. Iron is present in <u>excess</u>, so <u>chlorine</u> is limiting.

Now, calculate the theoretical yield, using the limiting reagent:

$$15.0\ g\ Cl_2\ \times\ \frac{1\ mole\ Cl_2}{70.906\ g\ Cl_2}\ \times\ \frac{2\ moles\ FeCl_3}{3\ moles\ Cl_2}\ \times\ \frac{162.206\ g\ FeCl_3}{1\ mole\ FeCl_3}\ =\ 22.9\ g\ FeCl_3.$$

The <u>yield</u> of a reaction is just the amount of product that is <u>actually</u> obtained from a reaction. The yield is determined by <u>experiment</u>; the theoretical yield is found by calculation. The <u>percentage yield</u> of a reaction is simply the <u>ratio</u> of the yield of a reaction to its theoretical yield, multiplied by 100%.

<u>Problem</u>: A student obtained a yield of 20.0 g of $FeCl_3$ when she reacted 10.0 g of Fe with 15.0 g of Cl_2. Calculate her percentage yield.

$$\underline{\text{Solution}}:\ \ \text{Percentage Yield} = \frac{\text{Yield} \times 100\%}{\text{Theoretical Yield}} = \frac{20.0\ g}{22.9\ g} \times 100\%$$

$$= 87.3\%\ \text{yield.}$$

Percentage yields of less than 100% are common. <u>Side reactions</u>, which give products other than the desired product, usually account for the "missing" mass. (A possible side reaction in this case is: $Fe\ +\ Cl_2\ \longrightarrow\ FeCl_2.$)

33

SOME TERMS USED TO DESCRIBE SOLUTIONS:

Solutions are homogeneous mixtures of two or more substances. Coffee, air, and brass are all examples of solutions. The substance which is present in the largest amount in a solution is usually called the solvent. Water is a common solvent (sometimes called the "universal solvent"), and solutions in which water is the solvent are called aqueous solutions (from the Latin word "aqua", meaning "water"). The substances which are dissolved in the solvent are called solutes. In soft drinks, the solutes include sugar, caffeine, and carbon dioxide.

The concentration of a solution is a description of the ratio of the amounts of solutes and solvents present. Qualitatively, solutions may be either dilute (solute/solvent ratio is relatively small) or concentrated (solute/solvent ratio is large). A solution which is so concentrated that no more solute will dissolve in it is called a saturated solution. However, it is sometimes possible to make a solution even more concentrated than this by heating the saturated solution (heating a solution usually increases the tendency of the solute to dissolve), adding more solute, stirring the solution to dissolve the solute, and carefully cooling the solution back to the original temperature. Solutions such as these are called supersaturated solutions.

Concentrations of solutions are usually expressed in quantitative terms for laboratory purposes. The most common of these expressions is the molarity of a solution, which is defined as the amount of solute present per liter of solution. "Amount of solute" is measured in moles, so concentrations expressed in this way are useful in the laboratory, where calculations involving moles are common. Molarity is abbreviated by "M". For example, a saturated solution of calcium hydroxide has a concentration of 0.0250 M, since 0.0250 moles of $Ca(OH)_2$ will dissolve in enough water to make 1.00 liter of solution at room temperature.

CALCULATIONS INVOLVING CONCENTRATIONS:

Problem: How would you make 250.0 mL of a 0.100 M solution of $AgNO_3$?

Solution: Since "M" means "moles per liter", multiply the volume of the solution (in liters) by the concentration to obtain the amount (in moles) of $AgNO_3$ needed. Converting this to a mass in grams, we obtain the following:

$$250.0 \text{ mL} \times \frac{1 \text{ L}}{1000 \text{ mL}} \times \frac{0.100 \text{ moles } AgNO_3}{1 \text{ L of solution}} \times \frac{169.873 \text{ g } AgNO_3}{1 \text{ mole } AgNO_3} = 4.25 \text{ g of } AgNO_3.$$

To actually make this solution, 4.25 g of $AgNO_3$ would be weighed out and placed in a 250-mL <u>volumetric</u> flask. (A volumetric flask is a vessel which has been calibrated to show the <u>exact</u> volume of liquid present in the flask.) Distilled water would then be added to the flask until the $AgNO_3$ dissolves and the total volume of the solution is exactly 250.0 mL. (This is <u>not</u> the same as adding 250.0 mL of water to the $AgNO_3$. Actually, <u>less</u> water would be required.)

Problem: How would you make 100.0 mL of a 2.90 M solution of HCl? (Solutions of HCl with a concentration of 11.6 M are commercially available.)

Solution: Determine the amount (in moles) of HCl present in the desired solution. Then, calculate the volume of the concentrated solution needed to provide this amount. The difference in volume will be made up by adding distilled water to make the concentrated solution more dilute. Hence:

$$100.0 \text{ mL} \times \frac{1 \text{ L}}{1000 \text{ mL}} \times \frac{2.90 \text{ moles HCl}}{1 \text{ L solution}} \times \frac{1 \text{ L solution}}{11.6 \text{ moles HCl}} \times \frac{1000 \text{ mL}}{1 \text{ L}}$$

$$= 25.0 \text{ mL of concentrated (11.6 M) HCl needed.}$$

To make this solution, add about 50 mL of distilled water to a 100-mL volumetric flask, then add 25.0 mL of 11.6 M HCl. (Acid should be poured into water, not the other way around -- if anything splashes, let it be water!) More distilled water should then be added to the volumetric flask to make the final volume exactly equal to 100.0 mL.

STOICHIOMETRY OF REACTIONS IN SOLUTION:

Problem: A scientist knocks over a volumetric flask containing 100.0 mL of a 2.90 M solution of HCl, spilling the solution on the lab bench. Since HCl is an acid, a base must be used to neutralize it, but the only base available is a 1.97 M solution of Na_2CO_3. What volume (in mL) of the Na_2CO_3 solution is needed to neutralize the HCl solution? The balanced equation of the reaction is:

$$2\ HCl_{(aq)}\ +\ Na_2CO_3{}_{(aq)}\ \longrightarrow\ 2\ NaCl_{(aq)}\ +\ H_2O_{(l)}\ +\ CO_2{}_{(g)}$$

where the abbreviations "aq", "l", and "g" stand for "aqueous solution", "liquid", and "gas", respectively. (Solids are indicated by the subscript "s".)

Solution: This problem is solved in the same way any stoichiometry problem must be solved -- namely, using moles. First, convert the known volume and concentration of HCl into an amount in moles:

$$100.0\ mL\ \times\ \frac{1\ L}{1000\ mL}\ \times\ \frac{2.90\ moles\ HCl}{1\ L\ solution}\ =\ 0.290\ moles\ of\ HCl.$$

Next, use the mole ratio of HCl to Na_2CO_3 to calculate the amount of Na_2CO_3 (in moles) needed to neutralize the above amount of HCl:

$$0.290\ moles\ of\ HCl\ \times\ \frac{1\ mole\ of\ Na_2CO_3}{2\ moles\ of\ HCl}\ =\ 0.145\ moles\ of\ Na_2CO_3\ needed.$$

Finally, use the concentration of Na_2CO_3 to convert this into the desired volume, expressed in mL:

$$0.145\ moles\ Na_2CO_3\ \times\ \frac{1\ L\ of\ solution}{1.97\ moles\ Na_2CO_3}\ \times\ \frac{1000\ mL}{1\ L}\ =\ 73.6\ mL\ of\ Na_2CO_3.$$

(Of course, had this been an actual emergency, the scientist should neutralize the acid spill first and do calculations later!)

Notice that in the above sequence of calculations, we divided by 1000 and multiplied by 1000. Since these two steps cancel each other out, they can be safely omitted by those who wish to work in millimoles. (mL x M = mmol!)

VOLUMETRIC ANALYSIS USING TITRATIONS:

Volumetric analysis refers to any kind of analytical method based on measuring volumes. A typical laboratory example is a titration, in which two solutions are combined until the reactants in each solution are exactly consumed.

The most common type of titration involves the neutralization of an acid by a base. Usually, a solution of a known base is made and standardized by titrating a known volume of the base with a known mass or volume of a known acid. The stoichiometry of the acid-base reaction is then used to calculate the concentration of the basic solution, which is then used to titrate an unknown acid. The results of the titration provide information about the unknown acid.

An indicator is used to determine the point at which the reactants in each solution are exactly consumed. (This point is called the end point, since the titration ends when this point is reached.) Indicators are substances which change color when some property of the solution in which they are dissolved changes. Typical indicators for acid-base titrations include phenolphthalein (which is colorless in acidic solutions but red in basic solutions) and bromothymol blue (which is yellow in acids but blue in bases). The appearance of an intermediate color (pink for phenolphthalein or green for bromothymol blue) signals the end point, at which the acid and base have been neutralized.

Problem: A 229-mg sample of an unknown acid was titrated using 29.70 mL of 0.0965 M NaOH solution. What is the molecular weight of the acid? The balanced equation of the reaction which occurs is: $HX + NaOH \longrightarrow NaX + H_2O$.

Solution: The necessary conversion factors are shown below.

$$29.70 \text{ mL NaOH} \times \frac{1 \text{ L}}{1000 \text{ mL}} \times \frac{0.0965 \text{ mol NaOH}}{1 \text{ L of solution}} \times \frac{1 \text{ mol HX}}{1 \text{ mol NaOH}} = 2.87 \times 10^{-3} \text{ moles of HX.}$$

$$\text{Molecular Weight} = \frac{229 \text{ mg of HX}}{2.87 \times 10^{-3} \text{ mol HX}} \times \frac{1 \text{ g}}{1000 \text{ mg}} = 79.9 \text{ g/mole.}$$

ENERGY AND ITS MEASUREMENT:

When you lift a heavy object, you oppose the force of gravity which pulls that object toward the center of the earth. This is one example of doing work. Work is done whenever a force is exerted against an opposing force. You are able to do work because you have energy. Energy is difficult to define, but you can think of it as something that an object has if it is able to do work.

Energy can have many forms, some of which should be familiar. For example, kinetic energy (sometimes called mechanical energy) is energy of motion, and is defined by the equation $KE = mv^2/2$, where m is an object's mass and v is its velocity. Potential energy is "stored" energy. For example, chemical energy is a form of potential energy which is stored within compounds in the form of attractive forces (called chemical bonds) which connect the atoms or ions of a compound to each other.

All of the different forms of energy can be totally converted into heat. Heat is a familiar form of energy -- you learned what "hot" and "cold" meant when you were a child! Heat is that form of energy which flows from "hot" objects toward "cold" objects. Heat is not the same as temperature -- temperature is a measurement of the intensity of the heat present in an object. Besides, the two are measured using different units: temperature is measured in degrees, but energy is measured in joules or calories. Since $KE = mv^2/2$, the units of energy must have the dimensions of mass x velocity^2. One joule = 1 kg m^2/sec^2. (An object with a mass of 2 kg moving at a velocity of 1 m/sec has a kinetic energy of 1 J.) One calorie = 4.184 joules. (This is a defined number, containing an infinite number of "sig figs"!) One calorie is also the amount of heat needed to raise the temperature of one gram of water by one Celsius degree. (The "calories" on food labels are actually kilocalories -- be aware of this!)

THERMOCHEMISTRY:

The potential energy of a system <u>increases</u> whenever objects which <u>repel</u> each other are brought closer together, or whenever objects which <u>attract</u> each other are separated. Compounds exist because the atoms or ions from which they are made attract each other. These attractive forces between atoms or ions in compounds are a form of potential energy called <u>chemical bonds</u>. Dalton's atomic theory states that atoms or ions are <u>rearranged</u> during a chemical reaction. As an example, consider the "thermite" reaction, which has been used for welding:

$$2\ Al\ +\ Fe_2O_3\ \longrightarrow\ 2\ Fe\ +\ Al_2O_3$$

Notice that in the course of this reaction, the oxygen atoms become separated from the iron atoms and end up attached to the aluminum atoms. One way to think of this as occurring is in a two-step process, as shown below:

$$Step\ One:\ 2\ Al\ +\ Fe_2O_3\ \longrightarrow\ 2\ Al\ +\ 2\ Fe\ +\ 3\ O$$
$$Step\ Two:\ 2\ Al\ +\ 2\ Fe\ +\ 3\ O\ \longrightarrow\ 2\ Fe\ +\ Al_2O_3$$

In the first step, the attractive forces between the Fe and O atoms are overcome. As these atoms are separated, the chemical bonds are "broken", and the potential energy of the system <u>increases</u>. In Step Two, new chemical bonds are formed due to the attraction of the Al atoms for the O atoms. As these atoms are brought closer together, the potential energy of the system <u>decreases</u>.

Changes in potential energy are usually observed as the consumption or release of <u>heat</u>. Therefore, the study of the changes in energy which occur during chemical reactions is called <u>thermochemistry</u> (from the Greek "therme", meaning "heat"). In Step One, the reaction <u>consumes</u> energy. Energy-consuming processes are called <u>endothermic</u> (from the Greek "endon", meaning "within"). In Step Two, the reaction <u>releases</u> energy. Energy-releasing processes are called <u>exothermic</u> (from the Greek "exo", meaning "out").

ENTHALPY OF REACTIONS:

The amount of heat released or consumed in a chemical reaction is called the <u>heat of reaction</u>, and is given the symbol q_r, where "q" means "heat" and "r" means "reaction". The value of q_r is a <u>positive</u> number for an <u>endothermic</u> (energy-consuming) process, and is a <u>negative</u> number for an <u>exothermic</u> (energy-releasing) process. Heats of reaction are measured in either joules or calories.

A chemical reaction which is carried out in a vessel that is open to the air is exposed to atmospheric pressure, which usually remains roughly constant over short periods of time. The <u>potential energy</u> of a system at <u>constant pressure</u> is called the <u>enthalpy</u> of the system. When a chemical reaction occurs, the potential energy (chemical energy) of the system changes, so the enthalpy changes. This <u>change in enthalpy</u> is simply the <u>heat of reaction</u>. In equation form, this is written: $\Delta H_r = q_r$, where "Δ" (the Greek letter <u>delta</u>) means "change" and "H" means "enthalpy".

We can tell whether a chemical reaction is exothermic or endothermic by determining whether it feels "hot" (gives off heat) or "cold" (consumes heat). That is, we observe <u>changes in temperature</u>. To measure temperature changes, all we need to know is the <u>initial</u> temperature and the <u>final</u> temperature of the reaction mixture, and then simply subtract the two: $\Delta T = T_{final} - T_{initial}$. Temperature is one of several quantities which are called <u>state functions</u>, since changes in temperature depend only on the initial and final <u>states</u> of the system being studied. <u>Enthalpy</u> is also a state function -- the enthalpy change of a reaction is simply the difference in enthalpy between the <u>products</u> (final state) and the <u>reactants</u> (initial state): $\Delta H_r = H_{products} - H_{reactants} = q_r$. This is true <u>regardless</u> of the actual route by which the reaction gets from the reactants to the products -- all that matters are the initial and final <u>states</u>.

STANDARD HEATS OF REACTIONS:

The amount of heat consumed or released by a given chemical reaction varies with the pressure and temperature at which the reaction occurs. Scientists who measure heats of reactions have chosen a set of <u>standard conditions</u> for measuring heats of reactions. These standard conditions are a temperature of 25.0 $^{\circ}$C (77.0 $^{\circ}$F, approximately "room temperature") and an atmospheric pressure of 1.0 <u>atmosphere</u> (29.92 inches of mercury, which are the units used to describe atmospheric pressure in weather reports).

The <u>standard heat of reaction</u> is simply the heat of reaction measured under the above standard conditions. A <u>thermochemical equation</u> is an equation of a reaction which also includes the standard heat of reaction. For example:

$$2\ Fe_{(s)} + \frac{3}{2}\ O_{2\ (g)} \longrightarrow Fe_2O_{3\ (s)} \qquad \Delta H_r^o = -196.5 \text{ kcal}^{[1]}$$

(The superscript "o" in the symbol "ΔH_r^o" indicates that this is a <u>standard</u> heat of reaction, measured at 25 $^{\circ}$C and a pressure of 1.0 atmosphere.)

Notice that under the standard conditions, iron is a solid and oxygen is a gas. These are the <u>most stable physical forms</u> of iron and oxygen at 25.0 $^{\circ}$C and 1.0 atmosphere. These elements are therefore said to be in their <u>standard states</u> -- that is, the physical states in which they would normally occur. The above reaction therefore represents the <u>formation</u> of solid Fe_2O_3 from the <u>elements</u> of which it is composed, in their standard states. The <u>standard heat of formation</u> of a compound is simply the heat of reaction for the reaction where the compound is formed from its elements in their standard states. The symbol "ΔH_f^o", where "f" means "formation", is used to indicate the standard heat of formation. Hence, the information in the above thermochemical equation can be summarized by saying that for $Fe_2O_{3\ (s)}$, $\Delta H_f^o = -196.5 \text{ kcal}.^{[1]}$

---

[1] CRC Handbook of Chemistry and Physics, 56th Edition, 1975-1976.

41

HESS' LAW:

Consider the _thermite_ reaction, which has been used for welding:

$$Fe_2O_3 (s) + 2\ Al (s) \longrightarrow Al_2O_3 (s) + 2\ Fe (s) \qquad \Delta H_r^O = -202.6\ kcal$$

Since enthalpy is a _state function_, the above value of H depends on just the initial and final states of the reaction, and not on the actual pathway by which the reaction gets from the reactants to the products. Therefore, we can mentally picture this reaction occurring by any pathway we choose. One possible reaction pathway is the decomposition of the reactants into the elements from which they are made, followed by the reaction of those elements to form products:

$$\text{Step One:}\quad Fe_2O_3 (s) \longrightarrow 2\ Fe (s) + \tfrac{3}{2} O_2 (g) \qquad \Delta H_r^O = +196.5\ kcal\,[1]$$

$$\text{Step Two:}\quad 2\ Al (s) + \tfrac{3}{2} O_2 (g) \longrightarrow Al_2O_3 (s) \qquad \Delta H_r^O = -399.09\ kcal\,[1]$$

Note that -202.6 kcal = (+196.5 kcal) + (-399.09 kcal). _Hess' Law_ states that for any chemical reaction that can be written as if it occurs in a stepwise fashion, the enthalpy change of the reaction equals the _sum_ of the enthalpy changes of the individual steps. In equation form, $\Delta H_r = \Delta H_{Step\ 1} + \Delta H_{Step\ 2} + \Delta H_{Step\ 3} + \ldots$

Step Two above is simply the equation of the formation of Al_2O_3 from its _elements_ in their _standard states_. Therefore, ΔH_r^O for this reaction is just the _standard heat of formation_ (ΔH_f^O) of Al_2O_3. Step One is similar, except the equation is _reversed_. ΔH_f^O for Fe_2O_3 is -196.5 kcal. From this we can see that reversing the reaction simply changes the _sign_ (+ or -) of the enthalpy change. Therefore, if the standard heats of formation of the reactants and products of a chemical reaction are available, then heats of reactions can be found using Hess' Law in its alternate form: $\Delta H_r^O = \Delta H_f^O\ (products) - \Delta H_f^O\ (reactants)$. This form is often useful, since values of ΔH_f^O have been compiled in reference texts.[1]

---

1) CRC Handbook of Chemistry and Physics, 56th Edition, 1975-1976.

CALCULATING STANDARD HEATS OF REACTIONS USING HESS' LAW:

The standard heats of formation of many compounds have been measured and recorded in reference texts.[1] Therefore, it is often possible to <u>calculate</u> the standard heat of reaction <u>without</u> doing the reaction. Examples appear below.

<u>Reaction</u>: $HCl_{(g)}$ + $NaOH_{(s)}$ $\longrightarrow$ $NaCl_{(s)}$ + $H_2O_{(l)}$

To calculate ΔH_r^O for this reaction, just look up the standard heats of formation of the reactants and products in reference tables, then apply Hess' Law in the form: $\Delta H_r^O = \Delta H_f^O$ (products) $- \Delta H_f^O$ (reactants). ΔH_f^O values[1]:

$HCl_{(g)}$ = -22.06 kcal/mole. $\qquad$ $NaCl_{(s)}$ = -98.23 kcal/mole.

$NaOH_{(s)}$ = -101.99 kcal/mole. $\qquad$ $H_2O_{(l)}$ = -68.32 kcal/mole.

(<u>Note</u>: Standard heats of formation are usually given as "kcal/<u>mole</u>" because the assumption is that <u>one mole</u> of the compound is being formed from its elements.)

$$\Delta H_r^O = \Delta H_f^O \text{ (products)} - \Delta H_f^O \text{ (reactants)}$$
$$= (-68.32 \text{ kcal/mole}) + (-98.23 \text{ kcal/mole})$$
$$- [(-22.06 \text{ kcal/mole}) + (-101.99 \text{ kcal/mole})] = -42.50 \frac{kcal}{mole}.$$

<u>Reaction</u>: $CH_{4\,(g)}$ + $2\,O_{2\,(g)}$ $\longrightarrow$ $CO_{2\,(g)}$ + $2\,H_2O_{(g)}$

The calculation is similar to the one above, except we now have some coefficients which are not equal to <u>one</u>. ΔH_f^O values must be multiplied by the coefficients in this case. Note also that the ΔH_f^O values for $H_2O_{(l)}$ and $H_2O_{(g)}$ are different. Also, the value of ΔH_f^O for $O_{2\,(g)}$ is <u>zero</u>. This makes sense -- oxygen is an <u>element</u>, and since <u>zero</u> heat is required to convert an element into itself, the standard heat of formation of <u>any element</u> is zero. ΔH_f^O values[1]:

$CH_{4\,(g)}$ = -17.889 kcal/mole. $\qquad$ $CO_{2\,(g)}$ = -94.052 kcal/mole.

$O_{2\,(g)}$ = 0.00 kcal/mole. $\qquad$ $H_2O_{(g)}$ = -57.80 kcal/mole.

$$\Delta H_r^O = (-94.052) + (2)(-57.80) - [(2)(0) + (-17.889)] = -191.76 \text{ kcal.}$$

---

1) CRC Handbook of Chemistry and Physics, 56th Edition, 1975-1976.

HESS' LAW AND THERMOCHEMICAL EQUATIONS:

Hess' Law states that if we can picture a chemical reaction as though it occurs in a stepwise fashion, then the enthalpy change for the reaction can be found simply by adding the enthalpy changes of the individual steps. <u>Any</u> sequence of steps which leads from the reactants to the products may be used. This is true because enthalpy is a <u>state function</u> -- the enthalpy change depends only on the initial and final <u>states</u> of the reaction, not on the way the reaction proceeds.

As an example of this process, consider the reaction below:

$$C_2H_2 \text{ (g)} + 2\, H_2 \text{ (g)} \longrightarrow C_2H_6 \text{ (g)} \qquad \Delta H^O_r = ?$$

One way in which we can picture this reaction as occurring is as a two-step process. In the first step, the C_2H_2 molecule reacts with one of the H_2 molecules to form a molecule of C_2H_4. In the second step, the other H_2 molecule reacts with the C_2H_4 molecule to form the C_2H_6 molecule. This sequence is summarized below:

$$\text{Step One:} \quad C_2H_2 \text{ (g)} + H_2 \text{ (g)} \longrightarrow C_2H_4 \text{ (g)} \qquad \Delta H^O_r = -41.698 \text{ kcal}[1]$$

$$\text{Step Two:} \quad C_2H_4 \text{ (g)} + H_2 \text{ (g)} \longrightarrow C_2H_6 \text{ (g)} \qquad \Delta H^O_r = -32.732 \text{ kcal}[1]$$

Simply adding these two thermochemical equations gives us a new thermochemical equation, which happens to be the same as the original chemical equation above:

$$C_2H_2 \text{ (g)} + 2\, H_2 \text{ (g)} \longrightarrow C_2H_6 \text{ (g)} \qquad \Delta H^O_r = -74.430 \text{ kcal}$$

Notice that in adding the two thermochemical equations together, all of the reactants were combined on the left side of the arrow, and all of the products were combined on the right side of the arrow. The result of this is:

$$C_2H_2 \text{ (g)} + H_2 \text{ (g)} + C_2H_4 \text{ (g)} + H_2 \text{ (g)} \longrightarrow C_2H_4 \text{ (g)} + C_2H_6 \text{ (g)}$$

To obtain the final equation, similar terms on the <u>same</u> side of the arrow were combined ($H_2 \text{ (g)} + H_2 \text{ (g)} = 2\, H_2 \text{ (g)}$), and similar terms on <u>opposite</u> sides of the arrow were discarded (the "reaction" $C_2H_4 \text{ (g)} \longrightarrow C_2H_4 \text{ (g)}$ has a $\Delta H^O_r = 0$).

---

1) CRC Handbook of Chemistry and Physics, 56th Edition, 1975-1976.

HEATS OF REACTION AND HESS' LAW:

Problem: Given the following thermochemical equations:

$$\tfrac{1}{2} \, N_2 \, (g) \; + \; O_2 \, (g) \; \longrightarrow \; NO_2 \, (g) \qquad \Delta H^o_r = + \, 8.1 \text{ kcal}[1]$$

$$4 \, NO_2 \, (g) \; + \; O_2 \, (g) \; \longrightarrow \; 2 \, N_2O_5 \, (g) \qquad \Delta H^o_r = -25.2 \text{ kcal}[1]$$

Calculate the standard heat of reaction for the reaction below:

$$N_2 \, (g) \; + \; \tfrac{5}{2} \, O_2 \, (g) \; \longrightarrow \; N_2O_5 \, (g) \qquad \Delta H^o_r = \, ?$$

Solution: Hess' Law allows us to manipulate thermochemical equations just like algebraic equations, as long as we do the same things to the values of ΔH^o_r that we do to the chemical equations. The difficult part of problems such as the one above is deciding _how_ to manipulate the thermochemical equations so that the result is the reaction whose enthalpy change we want to know. In the problem above, we want to manipulate the two thermochemical equations given so that when they are added, the resulting equation has N_2 and $\tfrac{5}{2} \, O_2$ among the reactants and N_2O_5 among the products. The first thermochemical equation given has $\tfrac{1}{2} \, N_2$ among the reactants, so it will be necessary to _multiply_ this equation (and its ΔH^o_r value) by _two_. The second thermochemical equation given has $2 \, N_2O_5$ among the products, so it will be necessary to _divide_ this equation (and its ΔH^o_r value) by two. Carrying out these operations gives the following thermochemical equations:

$$N_2 \, (g) \; + \; 2 \, O_2 \, (g) \; \longrightarrow \; 2 \, NO_2 \, (g) \qquad \Delta H^o_r = +16.2 \text{ kcal}$$

$$2 \, NO_2 \, (g) \; + \; \tfrac{1}{2} \, O_2 \, (g) \; \longrightarrow \; N_2O_5 \, (g) \qquad \Delta H^o_r = -12.6 \text{ kcal}$$

Now, simply adding these two equations, combining similar terms on the _same_ side of the arrow $(2 \, O_2 + \tfrac{1}{2} \, O_2 = \tfrac{5}{2} \, O_2)$, and canceling similar terms on _opposite_ sides of the arrow $(2 \, NO_2 \longrightarrow 2 \, NO_2)$ gives us the desired equation, with its ΔH^o_r:

$$N_2 \, (g) \; + \; \tfrac{5}{2} \, O_2 \, (g) \; \longrightarrow \; N_2O_5 \, (g) \qquad \Delta H^o_r = + \, 3.6 \text{ kcal}$$

$$(+16.2 + (-12.6))$$

---

1)   CRC Handbook of Chemistry and Physics, 56th Edition, 1975-1976.

HEATS OF FORMATION AND HESS' LAW:

Problem: Calculate ΔH_f^o for octane from the following information:

$$H_2\ (g)\ +\ \tfrac{1}{2}\ O_2\ (g)\ \longrightarrow\ H_2O\ (g) \qquad \Delta H_r^o = -57.80\ \text{kcal}\,[1]$$

$$C_{(s)}\ +\ O_2\ (g)\ \longrightarrow\ CO_2\ (g) \qquad \Delta H_r^o = -94.05\ \text{kcal}\,[1]$$

$$C_8H_{18}\ (1)\ +\ \tfrac{25}{2}\ O_2\ (g)\ \longrightarrow\ 8\ CO_2\ (g)\ +\ 9\ H_2O\ (g)$$

(octane, a component of gasoline) $\qquad \Delta H_r^o = -1219.03\ \text{kcal}\,[1]$

Solution: The heat of formation for octane is simply ΔH_r^o for the reaction in which octane is formed from its elements in their standard states. This reaction is: $8\ C_{(s)}\ +\ 9\ H_2\ (g)\ \longrightarrow\ C_8H_{18}\ (1)$. To arrive at this reaction, we need to manipulate the three equations given in the problem, along with their heats of reaction. Multiplying the first equation by 9 gives:

$$9\ H_2\ (g)\ +\ \tfrac{9}{2}\ O_2\ (g)\ \longrightarrow\ 9\ H_2O\ (g) \qquad \Delta H_r^o = -520.20\ \text{kcal}$$

Multiplying the second equation by 8 (to give us eight carbon atoms) gives:

$$8\ C_{(s)}\ +\ 8\ O_2\ (g)\ \longrightarrow\ 8\ CO_2\ (g) \qquad \Delta H_r^o = -752.40\ \text{kcal}$$

Reversing the third equation above simply changes the sign of ΔH_r^o:

$$8\ CO_2\ (g)\ +\ 9\ H_2O\ (g)\ \longrightarrow\ C_8H_{18}\ (1)\ +\ \tfrac{25}{2}\ O_2\ (g)$$

$$\Delta H_r^o = +1219.03\ \text{kcal}$$

Finally, adding these three new equations together, followed by combining similar terms on the same side of the arrow and canceling similar terms on opposite sides of the arrow, gives the desired equation, complete with the value of ΔH_r^o:

$$8\ C_{(s)}\ +\ 9\ H_2\ (g)\ \longrightarrow\ C_8H_{18}\ (1) \qquad \Delta H_r^o = -53.57\ \text{kcal}$$

$$((-520.20) + (-752.40) + (+1219.03) = -53.57) \qquad \Delta H_f^o = -53.57\ \text{kcal/mole}$$

This method illustrates one way to determine ΔH_f^o values in the lab -- measure the compound's heat of combustion (see the third equation in the problem), then apply Hess' Law to the heats of combustion of the compound and its elements.

---

1) CRC Handbook of Chemistry and Physics, 56th Edition, 1975-1976.

CALORIMETRY:

Changes in energy cannot be measured directly in the laboratory. What can be measured directly are the changes in temperature caused by exothermic or endothermic reactions. The science of measuring these temperature changes and using them to calculate the heats of chemical reactions (or the amounts of heat released or consumed by other processes) is called calorimetry.

All calorimetry experiments are based on the Law of Conservation of Energy, which states that energy is never created or destroyed, but simply changed from one form into another or transferred from one place to another. Simply put, if one object loses heat, another object must gain heat. Since the Law of Conservation of Energy (sometimes called the First Law of Thermodynamics) states that the total amount of energy present in the universe remains constant, the amount of heat lost by an object must equal the amount of heat gained by other objects. In equation form, this is written as: $q_{lost} = -q_{gained}$, where "q" means "heat" and the "-" is present because loss of heat is an exothermic process, for which q is negative, while gain of heat is an endothermic process, for which q is positive. The "-" sign just makes the other signs come out the way they should.

For precise experimental work, scientists often use an instrument called a "bomb" calorimeter. (It doesn't explode!) The "bomb" is a stainless steel vessel in which the reaction occurs. The "bomb" is usually sealed and immersed in an insulated tub of water, whose temperature changes are then noted. Heats of combustion are usually measured in a "bomb" calorimeter. For scientists on a limited budget, a styrofoam coffee cup provides reasonably good insulation against heat loss. "Coffee-cup" calorimeter experiments work well for reactions that take place in solution, giving reasonably good results. Precisely measuring temperature is important to the success or failure of a calorimetry experiment.

HEAT CAPACITY AND SPECIFIC HEAT:

We can convert temperature changes into amounts of energy if we know the heat capacity of the object whose temperature is changing. The <u>heat capacity</u> of an object is simply the amount of heat needed to raise the object's temperature by 1.00 ^{o}C. Heat capacity is measured in units of <u>energy per degree</u>, such as J/K or cal/^{o}C. Heat capacity is an <u>extensive</u> property of a substance, but <u>intensive</u> properties are generally more useful, so the "heat capacity per gram" of a substance is used frequently. This property is called the <u>specific heat</u> of a substance, and it is defined as the amount of heat needed to raise the temperature of 1.00 gram of a substance by 1.00 ^{o}C. Specific heat is measured in units of <u>energy per gram per degree</u>, such as $\dfrac{cal}{g\ K}$ or $\dfrac{J}{g\ ^{o}C}$.

<u>Problem</u>: A 70.0-gram piece of metal was heated to 77.0 ^{o}C and placed in 100.0 grams of water at 22.0 ^{o}C in a "coffee-cup" calorimeter. The metal and the water came to the same temperature at 27.0 ^{o}C. Calculate the heat capacity and the specific heat of the metal.

<u>Solution</u>: The first step is to figure out the amount of heat that was transferred from the metal to the water. This is easily done if you remember that the specific heat of water is 1.00 $\dfrac{cal}{g\ ^{o}C}$. (One calorie raises the temperature of one gram of water by one Celsius degree.) Since we know the mass of the water (100.0 g) and its temperature change ($\Delta T = T_f - T_i = 27.0\ ^{o}C - 22.0\ ^{o}C = 5.0\ ^{o}C$), we can calculate the amount of heat transferred by cancelling the proper units:

$$q_{water} = 1.00\ \frac{cal}{g\ ^{o}C} \times 100.0\ g \times 5.0\ ^{o}C = 500 \text{ calories.}\quad (\underline{Two}\text{ sig figs.})$$

Now, knowing that $q_{metal} = -q_{water} = -500$ calories, and $\Delta T_{metal} = T_f - T_i = 27.0\ ^{o}C - 77.0\ ^{o}C = -50.0\ ^{o}C$, we can calculate the above quantities as shown:

$$\text{Heat Capacity} = \frac{-500 \text{ cal}}{-50.0\ ^{o}C} \qquad\qquad \text{Specific Heat} = \frac{-500 \text{ calories}}{(70.0\ g)(-50.0\ ^{o}C)}$$

$$= 10 \text{ cal/}^{o}C.\quad (2 \text{ sig figs}) \qquad\qquad = 0.14 \text{ cal/g } ^{o}C.\quad (2 \text{ sig figs})$$

48

DETERMINING ENTHALPY CHANGES BY CALORIMETRY:

Problem: A sample of silver nitrate ($AgNO_3$) weighing 8.50 g was dissolved in 91.5 g of water in a "coffee-cup" calorimeter. The temperature of the water was 23.4 ^{o}C originally, but the addition of the $AgNO_3$ produced a drop in temperature of the solution, reaching a minimum temperature of 20.7 ^{o}C. Calculate the molar heat of solution of $AgNO_3$ -- that is, ΔH_r for the reaction: $AgNO_{3\ (s)} \longrightarrow AgNO_{3\ (aq)}$. Express your answer in kilocalories per mole of $AgNO_3$. Assume that the heat capacity of the calorimeter is zero and that the specific heat of the $AgNO_3$ solution is 1.00 $\frac{cal}{g\ ^{o}C}$.

Solution: In any calorimetry experiment, $q_{lost} = -q_{gained}$. In this experiment, the solution is losing heat (as seen by the drop in temperature), so the reaction must be gaining heat -- it is an endothermic reaction. Hence, the above equation can be rewritten: $q_{reaction} = -q_{solution}$. To determine the amount of heat transferred ($q_{solution}$), use the information supplied above:

Mass of Solution = 100.0 grams (8.50 g $AgNO_3$ + 91.5 g water)

$$\Delta T_{solution} = T_f - T_i = 20.7\ ^{o}C - 23.4\ ^{o}C = -2.7\ ^{o}C$$

$$q_{solution} = 1.00\ \frac{cal}{g\ ^{o}C} \times 100.0\ g \times (-2.7\ ^{o}C) = -270\ cal.\ \ (2\ sig\ figs.)$$

Therefore, $q_{reaction} = -q_{solution} = +270$ calories. To express this in kcal/mole, we need to calculate the amount of $AgNO_3$ (in moles) that was used.

$$8.50\ g\ AgNO_3 \times \frac{1\ mole\ AgNO_3}{169.873\ g\ AgNO_3} = 0.0500\ moles\ AgNO_3$$

$$\Delta H_r^{o} = \frac{+270\ calories}{0.0500\ moles\ AgNO_3} \times \frac{1\ kcal}{1000\ cal} = +5.4\ kcal/mole\ of\ AgNO_3.[1]$$

Note that the result is a positive number, which it should be for an endothermic process. The assumption that the heat capacity of the calorimeter is zero was made to simplify the calculations -- there was no "$q_{calorimeter}$" term present.

---

1) CRC Handbook of Chemistry and Physics, 56th Edition, 1975-1976.

CALORIMETRY EXPERIMENTS IN FOOD CHEMISTRY:

Problem: A teaspoonful of sugar (about 5.0 grams) was placed in a "bomb" calorimeter which had a heat capacity of 1.00 kJ/^{o}C. The calorimeter was immersed in a water bath containing 1.00 L of water. The sugar was combusted completely in the "bomb", causing the temperature of the water to rise from its initial value of 24.96 ^{o}C to a maximum value of 40.96 ^{o}C. Calculate the amount of energy released by this reaction, and compare your result to the advertiser's claim that sugar contains "only sixteen calories per teaspoon".

Solution: As with ary calorimetry problem, q_{lost} = -q_{gained}. In this case, the reaction releases heat, which is absorbed by the calorimeter and the water. Therefore, we can rewrite the above equation in the following way:

$$q_{reaction} = -[q_{water} + q_{calorimeter}]$$

Now, we can solve for the values of q_{water} and $q_{calorimeter}$ using the data above:

$$\Delta T_{water} = \Delta T_{calorimeter} = T_f - T_i = 40.96 \ ^{o}C - 24.96 \ ^{o}C = +16.00 \ ^{o}C.$$

$$q_{calorimeter} = (1.00 \ kJ/^{o}C) \times (+16.00 \ ^{o}C) \times \frac{1.00 \ kcal}{4.184 \ kJ} = +3.82 \ kcal.$$

$$q_{water} = 1.00 \ \frac{calorie}{gram \ ^{o}C} \times (+16.00 \ ^{o}C) \times 1.00 \ L \times 1000 \ mL/L \times 1.00 \ g/mL$$

$$= 16,000 \ cal = 16.0 \ kcal.$$

Finally, using the equation above, we can solve for $q_{reaction}$:

$$q_{reaction} = -[q_{water} + q_{calorimeter}] = -(16.0 \ kcal + 3.82 \ kcal)$$

$$= -19.8 \ kcal. \quad \text{(Three sig figs.)}$$

This is a little bit higher than the advertiser's claim of 16 "Calories" (food "Calories" are actually kilocalories) per teaspoon (level or heaping teaspoon?). This also works out to about 4.0 "Calories" per gram of sugar. This is a typical "Calorie content" (heat of combustion!) for a carbohydrate, which sugar is. (Sugar = $C_{12}H_{22}O_{11}$ = $C_{12}(H_2O)_{11}$ = "carbon-hydrate"!) Proteins also contain about 4.0 calories per gram, while fats contain over 9.0 calories per gram.

THE ORIGINS OF THE PERIODIC TABLE AND THE PERIODIC LAW:

Elements such as iron and sulfur have been known for millenia, but only in the last few centuries have scientists applied themselves to the study of the properties of the elements. By the middle of the 19th century, many of the more common elements had been studied and characterized, but this information was just a haphazard collection of observations -- it had not been organized in any particular way. In the 1860's, the Russian chemist Dmitri Mendeleev, while in the process of writing a chemistry textbook, wrote down the known information about the elements on index cards. He then noticed that if he arranged the cards in order of <u>increasing atomic weight</u> of the elements, the properties of the elements varied in a <u>periodic</u> way -- that is, every so often an element's properties would be similar to those of another element which had been previously encountered.

Mendeleev then arranged his index cards in a rectangular array, so that elements with similar properties were placed in the same column, and elements with different properties were placed in different columns. This arrangement left him with some "holes" in his <u>periodic table</u>, which he suspected corresponded to elements which had yet to be discovered. This hypothesis was shown to be correct; for example, the "hole" between silicon and tin was filled by germanium in 1886.

In 1912, the British physicist Henry Moseley discovered a method by which the <u>atomic numbers</u> (the numbers of protons or electrons present in atoms) of the elements could be accurately determined. Thereafter, it was found that a periodic table in which the elements were arranged in horizontal rows according to <u>increasing atomic number</u> was more consistent than Mendeleev's 1869 table. The modern version of the <u>periodic law</u> states that the chemical and physical properties of the elements vary in a <u>periodic</u> way as their <u>atomic numbers</u> increase. (Moseley died in battle during World War I at the age of 27.)

THE MODERN PERIODIC TABLE:

The Periodic Table of the Elements as we know it today is not quite the same as Mendeleev's periodic table. Today's Periodic Table consists of seven horizontal rows (sometimes called <u>periods</u>) numbered 1 to 7 from top to bottom. Each row contains between two and thirty-two elements. (The two rows which often appear at the bottom of the Periodic Table are actually part of Rows 6 and 7. The first of these rows, elements 58 through 71, are called the <u>lanthanides</u> after the 57th element, lanthanum. The other row, elements 90 through 103, are called the <u>actinides</u> after the 89th element, actinium.)

The elements are also arranged in vertical columns which are often called <u>groups</u>. Each group is designated by a Roman numeral and a letter. The groups which contain elements 1 through 20 are called the <u>main groups</u> and are assigned the letter "A". The groups which contain elements 21 through 30 are called the <u>transition metals</u> and are assigned the letter "B". The "A" groups are assigned the Roman numerals I through VIII going from left to right, but the "B" groups start with III on the left, increase through VIII (which describes the three-group block which contains elements 26-28, 44-46, and 76-78), and finishes with I and II. Thus, magnesium (Mg) is in Group IIA, chlorine (Cl) is in Group VIIA, chromium (Cr) is in Group VIB, and copper (Cu) is in Group IB.[1]

Some of the groups are referred to by their more common names. Thus, the Group IA elements are called the <u>alkali metals</u>, the Group IIA elements are called the <u>alkaline earth metals</u>, the Group VIIA elements are called the <u>halogens</u>, and the Group VIIIA elements are called the <u>noble gases</u>.

---

1) Author's Note: In 1986, the International Union of Pure and Applied Chemistry (IUPAC) proposed a new numbering system for the groups of the Periodic Table, consisting of the straightforward 1 through 18 from left to right. As this is being written, the controversy over the "best" way to number the groups remains unresolved. The system described above, while less straightforward than 1 through 18, retains certain advantages which will prove useful to students.

METALS AND NONMETALS:

The elements boron (B), silicon (Si), arsenic (As), tellurium (Te), and astatine (At) lie on a diagonal line in the Periodic Table. These elements and all elements which lie to the <u>right</u> of them on the Periodic Table are called <u>nonmetals</u>. Most of the elements which lie to the <u>left</u> of these on the Periodic Table are called <u>metals</u>. (The only exception is <u>hydrogen</u>, which is in Group IA but is a gas, not a metal!)

As you are probably aware, metals have certain properties in common which make them easy to recognize. Metals are <u>shiny</u>. In fact, the particular kind of shine common to most metals is so unique that it has been called the "metallic luster". Metals are <u>malleable</u> -- they can be molded into different shapes or drawn out into thin wires. This last property is especially useful, since metals are good <u>conductors</u> of heat and electricity. Most metals have <u>high</u> melting points and boiling points. (The best-known exception is <u>mercury</u> (Hg), which is a <u>liquid</u> at room temperature and is used in thermometers.)

Nonmetals are not quite so easy to characterize as a group, since their properties are so diverse. For example, the <u>halogens</u> (Group VIIA) include gases (fluorine and chlorine), a liquid (bromine), and a solid (iodine) among their ranks. Most nonmetals are <u>non-conductors</u> (sometimes called <u>insulators</u>) of heat and electricity. (A notable exception is carbon in its <u>graphite</u> form, which conducts electricity.) The solid nonmetals tend to be <u>brittle</u>, unlike the metals.

Some of the elements which lie near the "borderline" between metals and nonmetals on the Periodic Table have properties <u>intermediate</u> between those of metals and nonmetals. These elements are called <u>metalloids</u> or <u>semi-metals</u>, and they have been widely used in the <u>semiconductor</u> industry. Silicon, arsenic, tellurium, and germanium (Ge) are some of the elements which have these properties.

CHEMICAL PROPERTIES OF METALS AND NONMETALS:

$\underline{Ions}$ are formed when an atom gains or loses one or more electrons. As an example, consider a $\underline{hydrogen}$ atom, which has only $\underline{one}$ electron. Hydrogen atoms either lose their electrons to form H^+ ions, or gain electrons (one electron per atom) to form H^- ions. This is typical behavior for $\underline{nonmetals}$, whose atoms may either lose or gain electrons to form either $\underline{cations}$ (positive ions) or $\underline{anions}$ (negative ions). $\underline{Metals}$ do not have this choice. Atoms of metals always react by $\underline{losing}$ electrons, forming $\underline{positive}$ ions.

The number of electrons lost by a metal atom in forming an ion can be predicted from the Periodic Table. For the $\underline{main\text{-}group}$ metals ("A" in the group number), the number of electrons lost in forming an ion is simply the number of the group. Thus, $\underline{sodium}$ atoms (Group IA) form Na^+ ions, $\underline{calcium}$ atoms (Group IIA) from Ca^{2+} ions, and $\underline{aluminum}$ atoms (Group IIIA) from Al^{3+} ions. $\underline{Transition\ metals}$ ("B" in the group number) also form cations, but their chemistry is more difficult to predict because many transition metals may form $\underline{more\ than\ one\ cation}$. For example, $\underline{chromium}$ atoms (Group VIB) form Cr^{6+} and Cr^{3+} ions, among others.

The number of electrons gained by a nonmetal atom in forming an ion can also be predicted using the Periodic Table. Atoms of nonmetals tend to gain (8 - GN) electrons, where "GN" is the group number. The ions formed in this process have charges equal to (GN - 8). Thus, $\underline{chlorine}$ atoms (Group VIIA) gain $\underline{one}$ electron each (8 - 7 = 1) to form Cl^- ions (7 - 8 = -1). $\underline{Oxygen}$ atoms (Group VIA) gain $\underline{two}$ electrons each (8 - 6 = 2) to form O^{2-} ions (6 - 8 = -2). $\underline{Nitrogen}$ atoms (Group VA) gain $\underline{three}$ electrons each to form N^{3-} ions, and so on.

$\underline{Problem}$: Recalling that the positive and negative charges in a chemical formula must cancel each other, write formulas for the products when $\underline{hydrogen}$ reacts with: a) Ca, b) Al, c) Cl. ($\underline{Solutions}$: CaH_2, AlH_3, HCl.)

$$\text{THE ORIGINS OF SOME ACIDS AND BASES:}$$

Some <u>covalent</u> compounds form <u>ions</u> when dissolved in water. Examples include the covalent gases HCl and NH_3, as shown below:

$$HCl_{(g)} \;+\; H_2O_{(l)} \;\longrightarrow\; H_3O^+_{(aq)} \;+\; Cl^-_{(aq)}$$

$$NH_{3\,(g)} \;+\; H_2O_{(l)} \;\longrightarrow\; NH_4^+_{(aq)} \;+\; OH^-_{(aq)}$$

In 1903, the Swedish chemist Svante Arrhenius won the Nobel Prize in Chemistry for his studies of compounds such as HCl and NH_3 and their solutions. Arrhenius defined any substance which forms H_3O^+ ions (called <u>hydronium</u> ions) when dissolved in water as an <u>acid</u>, and any substance which forms OH^- ions (<u>hydroxide</u> ions) when dissolved in water as a <u>base</u>. By Arrhenius' definition, HCl is an acid (<u>hydrochloric acid</u>) and NH_3 (<u>ammonia</u>) is a base -- see the reactions above!

Some acids and bases can be formed by the reaction of various elements with oxygen and water vapor in the atmosphere. First, the element reacts with the oxygen to form a compound called an <u>oxide</u>. The oxide then reacts with the water vapor to form the acid or base.

Oxides of <u>metals</u> react with water to form <u>bases</u>. Examples include:

$$Na_2O_{(s)} \;+\; H_2O_{(l)} \;\longrightarrow\; 2\,NaOH_{(aq)}$$

$$CaO_{(s)} \;+\; H_2O_{(l)} \;\longrightarrow\; Ca(OH)_{2\,(aq)}$$

Oxides of <u>nonmetals</u> react with water to form <u>acids</u>. Examples include:

$$SO_{3\,(g)} \;+\; H_2O_{(l)} \;\longrightarrow\; H_2SO_{4\,(l)}$$

$$N_2O_{5\,(g)} \;+\; H_2O_{(l)} \;\longrightarrow\; 2\,HNO_{3\,(l)}$$

These last two reactions are important, because SO_3 and N_2O_5 are among the air pollutants emitted by some factories. When they react with water vapor in the atmosphere, the acids that form (H_2SO_4 = <u>sulfuric acid</u>, HNO_3 = <u>nitric acid</u>) return to earth in the form of <u>acid rain</u>. This causes severe environmental problems, such as fish kills and deterioration of forests.

RELATIVE STRENGTHS OF ACIDS AND THE PERIODIC TABLE:

When HCl reacts with H_2O, <u>all</u> of the HCl molecules dissociate (break
up) to form H_3O^+ and Cl^- ions. Acids such as HCl, which are <u>totally dissociated</u>
in water, are called <u>strong</u> acids. When HF reacts with H_2O, only about 3% of the
HF molecules dissociate to form H_3O^+ and F^- ions -- the other 97% remain as HF
molecules. Acids such as HF, which are only <u>partially dissociated</u> in water, are
called <u>weak</u> acids. (<u>Bases</u> can also be strong or weak. One mole of NaOH dissolved
in water will give one mole of OH^- ions, but one mole of NH_3 dissolved in water
will only give about 0.01 mole of OH^- ions. Thus, NH_3 is a <u>weak</u> base.)

The Periodic Table can be used to determine which of two acids is the
stronger acid. First, consider the formulas of the acids in question to determine
whether they are <u>binary acids</u> or <u>oxyacids</u>. <u>Binary acids</u> are composed of hydrogen
and <u>one other element</u> -- only <u>two</u> different elements are present in the formula.
<u>Oxyacids</u> are composed of hydrogen, <u>oxygen</u>, and one other element -- a total of
<u>three</u> different elements are present in the formula of an oxyacid.

The rule for determining the relative strengths of <u>binary acids</u> is
that the closer the "other" element is to the <u>lower right corner</u> of the Periodic
Table, the stronger the acid. HI is a stronger acid than HBr; I is closer to the
lower right corner than Br is. Similarly, HBr is a stronger acid than H_2Se.
(<u>Note</u>: Don't be fooled by the number of hydrogens present -- just use the rule!)

The rule for determining the relative strengths of <u>oxyacids</u> is that
the closer the "other" element is to the <u>upper right corner</u> of the Periodic Table,
the stronger the acid. H_2SO_4 is a stronger acid than H_2SeO_4; S is closer to the
upper right corner than Se is. Similarly, $HClO_4$ is a stronger acid than H_2SO_4.
(Don't let the number of hydrogens fool you!) The strength of an oxyacid <u>does</u>
increase as the number of <u>oxygens</u> present increases. HNO_3 is stronger than HNO_2.

ELECTROMAGNETIC ENERGY:

To understand more about the chemical behavior of atoms, we need to learn more about the configuration of the electrons around an atom's nucleus. Scientists have approached this problem by exposing atoms to various forms of electromagnetic energy. Since the first instruments used to do these experiments were called spectroscopes, the study of atoms using electromagnetic energy is called spectroscopy.

Electromagnetic energy consists of electrical and magnetic forces whose intensities oscillate as time elapses. You've seen something like this if you've ever thrown a stone into a still pond of water. The "ripples" which spread out from the point where the stone hit the water are called traveling waves. The position of anything on the surface of the water (such as a leaf or an insect) oscillates (moves up and down repeatedly) as the traveling waves pass by it.

The frequency of a wave of electromagnetic energy is simply the number of oscillations it undergoes in one second. Frequency is represented by the Greek letter nu (ν), and has units of cycles per second (an oscillation is sometimes called a "cycle"), also known as $\sec^{-1}$ or Hertz. The wavelength of a wave of electromagnetic energy is just the distance that the wave travels during one cycle. Wavelength is represented by the Greek letter lambda (λ), and has units of length or distance. Meters can be used, but often a more convenient unit is the nanometer. (The prefix "nano-" means 10^{-9}. One meter = 10^{9} nanometers.)

All electromagnetic energy travels at the speed of light, which is represented by the letter c and is equal to 3.00×10^{8} meters per second. Hence, the product of the frequency and the wavelength of a beam of electromagnetic energy is always equal to the speed of light: 3.00×10^{8} m/sec = c = $\lambda\nu$. (Bad Joke: First Physicist: "What's Nu?" Second Physicist: "C over Lambda!")

$$\text{CONTINUOUS AND DISCONTINUOUS SPECTRA:}$$

The many forms of electromagnetic energy make up a <u>continuous</u> <u>spectrum</u> -- <u>all</u> possible wavelengths and frequencies are represented below.

E.M. Energy	Wavelength, meters	Frequency, Hertz
Radio Waves	3×10^4 - 30	10^4 - 10^7
TV Waves	30 - 3	10^7 - 10^8
Radar	3 - 0.3	10^8 - 10^9
Microwaves	0.3 - 3×10^{-4}	10^9 - 10^{12}
Infrared	3×10^{-4} - 7×10^{-7}	10^{12} - 4×10^{14}
Visible Light	7×10^{-7} - 4×10^{-7}	4×10^{14} - 7×10^{14}
Ultraviolet	4×10^{-7} - 3×10^{-8}	7×10^{14} - 10^{16}
X Rays	3×10^{-8} - 3×10^{-11}	10^{16} - 10^{19}
Gamma Rays	3×10^{-11} - 3×10^{-16}	10^{19} - 10^{24}

A more common example of a continuous spectrum is the "rainbow" seen when white light is separated into its component colors by a prism. Each color "blends into" the next; there are no "gaps" in the spectrum. However, if a sample of hydrogen gas receives a high-energy electric spark, the H_2 molecules break apart into individual hydrogen atoms, which then get rid of their excess energy in the form of reddish-purple light. If the light emitted from these hydrogen atoms is passed through a prism, only the following wavelengths of visible light are present: 656 nm (red), 486 nm (green), 434 nm (indigo), and 410 nm (violet). The absence of the other wavelengths of visible light explains why the hydrogen spectrum is referred to as a <u>discontinuous spectrum</u> or a <u>line spectrum</u> -- there <u>are</u> "gaps" in the spectrum; only a <u>few</u> wavelengths are present. The hydrogen spectrum was discovered by Balmer in 1885. The spectra produced by other elements are also discontinuous. Scientists sought an explanation for this phenomenon.

PLANCK'S QUANTUM THEORY AND THE PHOTOELECTRIC EFFECT:

Hot objects glow. The hotter they get, the more the color of the glow moves toward the _blue_ end of the visible spectrum. Hot objects glow _red_, hotter objects glow _orange_, really hot objects glow _white_, and the hottest stars glow _bluish-white_. Classical physics predicts that this trend should continue indefinitely, with _extremely_ hot objects giving off _ultraviolet_ light as their primary emission. They don't. This situation was called the "ultraviolet catastrophe", because classical physics failed to predict the trend correctly.

In 1900, the German physicist Max Planck found a solution to this problem. The solution was based on the idea that energy comes in small particles which Planck called _photons_ or _quanta_, and that the energy of a quantum is related to its _frequency_. In equation form, Planck's _quantum theory_ is: $E_{photon} = h\nu$. Here, E_{photon} is the energy of a photon in joules, ν is its frequency in sec^{-1}, and h is _Planck's Constant_, which has the value: $h = 6.63 \times 10^{-34}$ joule seconds. Planck was awarded the 1918 Nobel Prize in Physics for this work.

The quantum theory received some support in 1905 from another German physicist -- Albert Einstein. Einstein was studying the _photoelectric effect_, which was first observed in 1887 by the German physicist Heinrich Hertz. Hertz observed that shining electromagnetic energy on some metals caused them to lose electrons from their surfaces. Furthermore, the _intensity_ of the energy used didn't matter -- what _did_ matter was that the energy had to have a certain minimum _frequency_ (called the _threshhold frequency_) for this effect to occur. Einstein explained this phenomenon using the quantum theory. If the energy of a quantum is proportional to its frequency, then only photons with a certain minimum _frequency_ (or a greater frequency) will have enough _energy_ to produce this effect. For this work (_not_ the Theory of Relativity!), Einstein won a Physics Nobel Prize in 1921.

BOHR'S "SOLAR SYSTEM" MODEL OF THE ATOM:

In 1913, the Danish physicist Niels Bohr proposed a "solar system" model of the atom to try to explain the line spectrum obtained by the emission of light from excited hydrogen atoms. According to Bohr's model, electrons were restricted to certain orbits (also called energy levels) surrounding the atomic nucleus -- they could not exist between these orbits. (A good analogy to this situation is someone standing on a staircase. S/he can be standing on the first step, the second step, or the third step, but not on the "$1\frac{1}{2}$th" step or the "$2\frac{1}{4}$th" step -- no such step exists!) Bohr described the energies of the energy levels by the equation: $E_{el} = -k/n^2$, where E_{el} is the energy of an energy level (or of an electron in that energy level), k is a constant with a value of 2.18 x 10^{-18} J, and n is just the number of the energy level. The energy levels are consecutively numbered, with the closest orbit to the nucleus having n = 1, the next orbit out having n = 2, the next orbit having n = 3, the next having n = 4, and so on.

Bohr postulated that when an atom absorbs electromagnetic energy, it stores that energy, by moving its electrons from the ground state (the energy level where n = 1 and E_{el} is a minimum) to higher-energy excited states (energy levels where n = 2 or more and E_{el} is greater than -k). When an atom emits electromagnetic energy, the reverse process occurs -- electrons in high-energy excited states move into lower-energy states.

Bohr's model also supports Planck's quantum theory. If the electrons in an atom can exist only in certain energy levels, and if they move from one level to another when the atom absorbs energy, then it must be true that the atom can only absorb certain amounts of energy. This is just another way of saying that energy is quantized -- that is, it comes in discrete "lumps" called quanta.

Bohr was awarded the 1922 Nobel Prize in Physics for this work.

USING BOHR'S MODEL TO PREDICT THE HYDROGEN EMISSION SPECTRUM:

The major significance of the Bohr model is that it allows for the prediction of the discontinuous spectrum of visible light emitted by excited atoms of hydrogen. The hydrogen spectrum consists of visible light with the following wavelengths: 656 nm, 486 nm, 434 nm, and 410 nm. Does the model predict this?

Problem: What is the wavelength of the photon of light emitted when an electron moves from the n = 3 level to the n = 2 level in the hydrogen atom?

Solution: When an electron moves from one energy level to another, the energy of the electron changes. The Law of Conservation of Energy tells us that the electron's change in energy must be equal to the energy of the photon of light that is emitted. Thus, we need to find the energies of the energy levels involved -- here, the n = 3 and n = 2 energy levels. Given that $E_{el} = -k/n^2$ and knowing that $k = 2.18 \times 10^{-18}$ joules, we can calculate the following:

$$E_3 = -2.18 \times 10^{-18} \text{ J}/3^2 = -2.18 \times 10^{-18} \text{ J}/9 = -2.42 \times 10^{-19} \text{ joules.}$$

$$E_2 = -2.18 \times 10^{-18} \text{ J}/2^2 = -2.18 \times 10^{-18} \text{ J}/4 = -5.45 \times 10^{-19} \text{ joules.}$$

Now, simply subtracting these two values gives us the energy of the photon:

$$E_{photon} = E_3 - E_2 = (-2.42 \times 10^{-19} \text{ joules}) - (-5.45 \times 10^{-19} \text{ joules})$$
$$= 3.03 \times 10^{-19} \text{ joules.}$$

Planck's equation ($E_{photon} = h\nu$) can be used to convert this into a frequency:

$$\nu = E_{photon}/h = (3.03 \times 10^{-19} \text{ joules})/(6.63 \times 10^{-34} \text{ joule sec})$$
$$= 4.57 \times 10^{14} \text{ sec}^{-1}.$$

Finally, the wavelength can be found using the $c = \lambda\nu$ relationship:

$$\lambda = c/\nu = (3.00 \times 10^8 \text{ meters/sec})/(4.57 \times 10^{14} \text{ sec}^{-1})$$
$$= 6.56 \times 10^{-7} \text{ meters} = 656 \times 10^{-9} \text{ meters} = \underline{656 \text{ nanometers!}}$$

Notice that this is one of the wavelengths observed in the hydrogen spectrum! The other wavelengths above can be calculated in a similar way. See if you can do it!

$$\text{THE WAVE NATURE OF MATTER:}$$

Bohr's "solar system" model of the atom turned out to be useful <u>only</u> for predicting the emission spectrum of <u>hydrogen</u>. For other elements, the Bohr model failed miserably, so another model was needed for more complex atoms.

In 1924, a French graduate student in physics, Louis deBroglie, combined Planck's equation ($E = h\nu$) and Einstein's famous relationship between mass and energy ($E = mc^2$) into one equation: $h\nu = mc^2$. Dividing both sides of this equation by $mc\nu$ gives us a new equation: $\dfrac{h}{mc} = \dfrac{c}{\nu} = \lambda$. (since $c = \lambda\nu$) deBroglie's equation, for which he won the 1929 Nobel Prize in Physics, shows that <u>matter</u> (represented by m in the equation) has <u>wave</u> properties (represented by λ).

Following the publication of deBroglie's (four-page!) Ph.D. thesis which included the above hypothesis, numerous experiments supported the idea that <u>wave</u> properties are exhibited by some forms of <u>matter</u>. For example, in 1927, physicists Clinton Davisson and George Thomson (son of J. J. Thomson, who discovered electrons!) found that a beam of electrons passing through a crystal was scattered in such a way as to produce a "diffraction pattern" -- a series of alternating bright spots and dark spots which occurs when <u>waves</u> reinforce each other or cancel each other out, respectively. (The analogy of two stones thrown into a still pond may be useful in thinking about what a "diffraction pattern" is. As the "ripples" spread out from each stone, they will eventually overlap each other. The pattern of waves that forms when they overlap is the "diffraction pattern" for these waves.) Since electrons behave like <u>waves</u> in this experiment, scientists began to look for <u>wave</u> models to describe the behavior of electrons in atoms. Davisson and Thomson shared the 1937 Nobel Prize in Physics for this work. (Ironically, J. J. Thomson showed that electrons were <u>particles</u>, while his son George showed that electrons were <u>waves</u>! Both men deserved their Nobel Prizes.)

WAVE EQUATIONS FOR ELECTRONS:

Beams of electrons behave like traveling waves (that is, like ripples on a still pond when a stone is thrown into it), but electrons in atoms aren't traveling in a straight line. Instead, they tend to stay relatively close to the atomic nucleus. Electrons in atoms are thought to behave more like standing waves. A common example of a standing wave is a vibrating guitar string. The string obviously oscillates back and forth, but it doesn't travel anywhere as long as the guitar itself stands still. Different wave patterns can be produced by holding down one or more points along the guitar string before plucking it, but the wave patterns are limited by the fact that the ends of the guitar string are already tied to the guitar. The wave equation which describes the wavelength of a standing wave is: $\lambda = 2L/n$, where λ is the wavelength, L is the length of the guitar string, and n is a number related to the number of nodes present in the standing wave. (A node is just a part of a standing wave that doesn't move -- in the guitar string, a part that's held down.) The point is that only certain values of the wavelength are possible for a standing wave -- just as in the Bohr model of the atom, only certain orbits were possible! In each case, n may have the values 1, 2, 3, 4, etc., but no values between these numbers are allowed.

In 1926, the Austrian physicist Erwin Schrödinger worked out the wave equations which describe the probability of finding electrons at various locations in the atom. We can only discuss the probability of finding an electron at a specific location, because any attempt to measure its exact location causes it to move. This was discovered in 1926 by the German physicist Werner Heisenberg, who stated the Heisenberg Uncertainty Principle this way: "It is impossible to know both the position and the momentum of a small particle simultaneously." Both Heisenberg (1932) and Schrödinger (1933) received Nobel Prizes in Physics.

ATOMIC ORBITALS:

The solutions to Schrödinger's wave equations for electrons in atoms are very complex, but they can be graphed to see what they look like. Actually, drawing a graph on paper is somewhat misleading -- paper is two-dimensional, but the solutions to the Schrödinger ecuations represent three-dimensional regions called orbitals. An orbital is simply a region in three-dimensional space in which electrons in atoms are most likely to be found. Orbitals are centered at the nucleus of an atom. They are called "orbitals" because they are similar to (but not the same as!) the "orbits" for electrons in the Bohr model of the atom. For example, just as the different "orbits" in the Bohr model had different energies, different orbitals also have different energies. When an atom absorbs energy or emits energy, the atom's electrons move from their original orbitals to other orbitals with different energies. However, since orbitals are just the graphs of wave equations, they have some properties similar to those of waves -- particularly, standing waves. For example, orbitals have different shapes, which are designated by different letters and tell us where electrons are most probably located. s orbitals are spherical, with the atomic nucleus located at the center of the sphere. s orbitals contain no nodes. p orbitals are shaped like "dumbbells". There are three kinds of p orbitals, arranged at 90^{0} angles to each other along the x, y, and z axes. p orbitals contain one node each. d orbitals come in groups of five. Four of them look like "four-leaf clovers"; the other one looks like a p orbital inside a "doughnut". d orbitals have two nodes each. f orbitals come in groups of seven. They are by far the most exotic-looking: three of them look like p orbitals inside two "doughnuts", and the other four look like "eight-leaf clovers", if you can imagine such a thing! f orbitals have three nodes. (We won't use f orbitals often, but we will use s, p, and d orbitals.)

QUANTUM NUMBERS:

The equation for the energy of an electron in the Bohr model of the atom is: $E_{el} = -k/n^2$, where n is a positive integer (1, 2, 3, 4, etc.) that represents the energy level in which the electron is located. The Bohr model worked well for hydrogen atoms, which only contain one electron each. However, a more complex equation (Schrödinger's) is needed to describe more complex atoms. Numbers such as n in the equation above are called quantum numbers, because they are quantized -- that is, they are only allowed to have certain values. The complete solution to Schrödinger's equation for complex atoms (those with more than one electron) requires four quantum numbers, which are described below.

The principal quantum number (symbolized by n) is similar to the "n" in the Bohr equation. It must be a positive integer (1, 2, 3, 4, etc.). As the value of n increases, the energy of the electron increases, and the distance of the electron from the atomic nucleus increases. (Just like the Bohr model!)

The azimuthal quantum number (symbolized by l) ranges in value from zero up to n-1, by integers: l = 0, 1, 2, 3, ..., n-1. The value of l designates the shape of the orbital in which the electron is located: l = 0 for an s orbital, l = 1 for a p orbital, l = 2 for a d orbital, and l = 3 for an f orbital.

The magnetic quantum number (symbolized by m) ranges in value from -l to +l, by integers. For example, if l = 2 (a d orbital), then the allowed values of m are -2, -1, 0, +1, and +2 -- a total of five different values of m, which tells us that there are five different d orbitals.

The spin quantum number (symbolized by s) has only two possible values: s = +1/2 or s = -1/2. Both the electron and the atomic nucleus are surrounded by magnetic fields, and the value of s tells us whether these magnetic fields point in the same direction ("spin up") or in opposite directions ("spin down").

THE PAULI EXCLUSION PRINCIPLE:

In 1926, the Austrian physicist Wolfgang Pauli discovered that no two electrons in the same atom can have the same values for all four quantum numbers. This axiom, often called the Pauli Exclusion Principle, is useful because it allows us to use sets of quantum numbers like "zip codes" for electrons in atoms. No two cities have the same zip code, and no two electrons have the same "address" in the atom. Notice that since the first three quantum numbers (n, l, and m) refer to orbitals, and only the spin quantum number (s) refers specifically to electrons, a corollary of the Pauli Exclusion Principle is that there can be a maximum of two electrons in any one orbital. Furthermore, these two electrons must be "spin-paired" -- that is, one of them must be "spin up" (s = +1/2) and the other one must be "spin down" (s = -1/2). Single electrons in orbitals are sometimes called "unpaired" electrons as a result. Pauli won the 1945 Nobel Prize in Physics for this discovery.

All of the orbitals which have the same value for the n quantum number are referred to as a shell, after the spherical, "shell-like" orbits of the Bohr model of the atom. Notice that in the first shell (n = 1), only one orbital is present (the 1s orbital), so that the maximum number of electrons that can be present in the first shell is two. In the second shell (n = 2), four orbitals are present (the 2s orbital and the three 2p orbitals), so a maximum of eight electrons may be present in the second shell. In the third shell (n = 3), nine orbitals are present (3s, three 3p, and five 3d orbitals), so at most eighteen electrons may be present. By similar logic, thirty-two electrons is the most that can be present in the fourth shell. Now, look at the Periodic Table. There are two elements in the first row, eight elements each in Rows 2 and 3, eighteen in Rows 4 and 5, and thirty-two in rows 6 and 7 -- just as Pauli's principle predicts!

THE AUFBAU PRINCIPLE:

The Aufbau principle ("Aufbau" comes from German words which mean "to build up") states that electrons in atoms will occupy the orbitals with the lowest energy first, and only after these orbitals are filled will electrons occupy orbitals of higher energy. The energies of orbitals increase as the values of the first two quantum numbers (n and l) increase. Thus, the 1s orbital is lowest in energy, the 2s orbital is next lowest, then the 2p orbitals, the 3s orbital, the 3p orbitals, and so on in the following sequence: 4s, 3d, 4p, 5s, 4d, 5p, 6s, 4f, 5d, 6p, 7s, 5f, 6d, 7p. An easy way to remember this sequence is to construct a "Pascal's Christmas Tree"[1], which is illustrated at the right.

```
            1s
            2s
         2p  3s
         3p  4s
      3d  4p  5s
      4d  5p  6s
   4f  5d  6p  7s
   5f  6d  7p  8s
            !!
```

Draw an eight-layer "Christmas tree" with one "bulb" each in rows 1 and 2, two "bulbs" each in rows 3 and 4, three "bulbs" each in rows 5 and 6, and four "bulbs" each in rows 7 and 8. Write an "s" in the right-hand bulb in each row, a "p" to the left of each "s", a "d" to the left of each "p", and an "f" to the left of each "d". Number the "s" bulbs from 1 to 8 (top to bottom), the "p" bulbs from 2 to 7, the "d" bulbs from 3 to 6, and the "f" bulbs from 4 to 5. Simply reading the rows across, from left to right, will give the above sequence of orbitals.

All of the orbitals which have the same values for both the n and l quantum numbers are referred to as a subshell. All of the orbitals in a subshell have the same energy, and are referred to as degenerate orbitals. When electrons are placed in a set of degenerate orbitals, they tend to "spread out" as much as possible -- one electron per orbital unless there are so many electrons that they must "pair up". This is known as Hund's Rule, sometimes called the "bus-seat" rule, because strangers boarding buses don't sit next to anyone unless they must!

---

1) Darsey, J. A., J. Chem. Educ. 65, 1036 (1988).

ELECTRON CONFIGURATIONS IN ATOMS:

Problem: What is the configuration of electrons in an _oxygen_ atom?

Solution: Oxygen's atomic number is $\underline{8}$, so a neutral oxygen atom will contain _eight_ electrons. The 1s orbital is lowest in energy, so we fill it first by placing _two_ electrons in it. The next two electrons go into the 2s orbital, since it is next lowest in energy. Next come the three degenerate 2p orbitals, but we only have _four_ electrons remaining. At this point, Hund's Rule says that we should place _one_ electron in _each_ of the three 2p orbitals, followed by placing the last electron in _any_ of the 2p orbitals to complete an electron pair. The resulting electron configuration is shown in the energy-level diagram at the right, but a more convenient "shorthand" way of representing the electron configuration of an atom is to simply list the subshells which contain electrons and write the number of electrons in each subshell as superscripts. By this method, oxygen's electron configuration is: $\underline{1s^2 2s^2 2p^4}$.

Problem: What is the electron configuration in a _tungsten_ atom?

Solution: Tungsten's atomic number is $\underline{74}$, so we need to place 74 electrons in orbitals. An energy-level diagram like the one above for this many electrons would be a tedious thing to draw, so we'll use the "shorthand" notation instead. All we need to do is remember the _order_ in which the subshells fill up, and if we need help with that, we can make a "Pascal's Christmas Tree"[1] (see the previous page). Then, if we know how many electrons can go into each subshell ($s = 2$, $p = 6$, $d = 10$, $f = 14$), all we have to do is count to 74 electrons! The correct result is: $\underline{1s^2 2s^2 2p^6 3s^2 3p^6 4s^2 3d^{10} 4p^6 5s^2 4d^{10} 5p^6 6s^2 4f^{14} 5d^4}$. This method will usually give the correct electron configuration, although there are some exceptions to the general rule (most of them fifth-row transition metals).

---

1)   Darsey, J. A., _J. Chem. Educ._ **65**, 1036 (1988).

ELECTRON CONFIGURATIONS AND THE PERIODIC TABLE:

The <u>valence shell</u> of an atom is the shell <u>farthest</u> from the nucleus which contains at least one electron. Many of the chemical properties of an atom are based on the number of <u>valence electrons</u> (electrons in the valence shell) that the atom possesses.

Problem: How many valence electrons does an atom of <u>sulfur</u> have?

Solution: Following the usual rules for determining electron configurations, we would obtain the following electron configuration for sulfur (atomic number = 16): $\underline{1s^2 2s^2 2p^6 3s^2 3p^4}$. At this point, we can easily see that the valence shell is the <u>third</u> shell (the largest value of n written above is 3), and that this shell contains <u>six</u> electrons (two in the 3s subshell and four in the 3p subshell). Hence, there are <u>six</u> valence electrons present in an atom of sulfur. However, this method would be tedious for atoms with large atomic numbers. There is a much easier way to obtain the above information -- use the Periodic Table! Sulfur is in the <u>third</u> row and in Group <u>VIA</u>, and it has <u>six</u> valence electrons in the <u>third</u> shell. In general, for any of the "main-group" elements ("A" groups), the <u>row number</u> is the same as the value of <u>n</u> for the <u>valence shell</u>, and the <u>group number</u> is the number of <u>valence electrons</u> present in the atom.

Another way to write the above electron configuration is: $\underline{[Ne]3s^2 3p^4}$. This "noble gas" form for writing electron configurations uses the symbol for a noble gas as "shorthand" for that gas ([Ne] = neon = $1s^2 2s^2 2p^6$) and emphasizes only the <u>valence</u> electrons by writing them outside the brackets.

Problem: What is the electron configuration for <u>calcium</u> (Ca) using the "noble gas" notation?

Solution: Calcium is in Row 4 and Group IIA. The noble gas with a filled third shell is <u>argon</u>, so the electron configuration is simply: $\underline{[Ar]4s^2}$.

LEWIS SYMBOLS AND THE OCTET RULE:

When metals react with nonmetals to form ionic compounds, electrons are transferred *from* the atoms of the metal *to* the atoms of the nonmetal. The cations and anions that result are then attracted to each other. This attraction is called an <u>ionic bond</u>. One way to keep track of the transfer of electrons that occurs in reactions like this is to use <u>Lewis symbols</u>, which were developed by the American chemist G. N. Lewis. A Lewis symbol is simply the chemical symbol for an element surrounded by <u>dots</u> which represent the <u>valence electrons</u> in the atoms of the element. Some examples of Lewis symbols are shown below.

$$Na = [Ne]3s^1 = Na\cdot \qquad\qquad P = [Ne]3s^2 3p^3 = \:\dot{\underset{\cdot}{\ddot{P}}}\cdot$$

$$Cl = [Ne]3s^2 3p^5 = \:\ddot{\underset{\cdot\cdot}{Cl}}\cdot \qquad\qquad Ar = [Ne]3s^2 3p^6 = \:\ddot{\underset{\cdot\cdot}{Ar}}\:$$

Note that the arrangement of the dots corresponds to the arrangement of the electrons in the valence shell in each case. Lewis symbols are sometimes called "electron dot" symbols, for obvious reasons.

An example of the use of Lewis symbols to illustrate a reaction in which electrons are transferred from one atom to another is shown below by the reaction of a calcium atom ($Ca = [Ar]4s^2$) with a sulfur atom ($S = [Ne]3s^2 3p^4$):

$$Ca\colon \; + \; \:\ddot{S}\colon \;\longrightarrow\; Ca^{2+} \; + \; \:\ddot{\underset{\cdot\cdot}{S}}\colon^{2-}$$

Notice that the calcium atom <u>loses</u> two electrons, so that the Ca^{2+} ion has the same electron configuration as an argon atom -- that is, [Ar]. Notice also that the sulfur atom <u>gains</u> two electrons, so that the S^{2-} ion also has the [Ar] configuration. The atoms of the main-group elements generally tend to gain or lose electrons until the ions formed by this process have the same electron configuration as the atoms of a <u>noble gas</u>. This is sometimes referred to as the <u>octet rule</u>, since the products of these reactions are usually ions or atoms with <u>eight</u> valence electrons. (An "octet" is <u>eight</u> of something.)

COVALENT BONDS:

Covalent compounds are made of <u>molecules</u>. The atoms in molecules are connected by <u>covalent bonds</u>. The word "co-valent" implies that these bonds are made of electrons which occupy the <u>valence</u> shells of <u>two</u> atoms simultaneously -- that is, that the electrons in a covalent bond are "shared" between two atoms.

In order to understand this better, consider two atoms, X and Y. Each atom has <u>one</u> unpaired electron. As the atoms move closer together, each unpaired electron begins to be attracted to the nucleus of the <u>opposite</u> atom, as well as to its own atom's nucleus. The covalent bond is fully formed when each electron is equally attracted to <u>each</u> of the two nuclei. An equation for this is shown below:

$$X\cdot \ + \ \cdot Y \ \longrightarrow \ X{:}Y$$

Covalent bonds can be represented by <u>dots</u> (for individual electrons) or by <u>dashes</u> (for <u>pairs</u> of electrons). For example, a molecule of H_2 could be represented by either H:H or H-H. Notice that the dots or dashes are drawn <u>between</u> the connected atoms, since the electrons in covalent bonds are located (roughly) between the two atomic nuclei that they join. Representations of covalent molecules like H:H or H-H are called <u>Lewis structures</u>, after the American chemist G. N. Lewis.

The covalent bond in H_2 is called a <u>single</u> bond, since it is composed of only <u>one</u> pair of electrons. <u>Multiple</u> bonds, consisting of two or more pairs of electrons, are also known to exist. Examples include the two <u>double</u> bonds in each molecule of CO_2 ($:\overset{..}{O}::C::\overset{..}{O}:$ or $:\overset{..}{O}{=}C{=}\overset{..}{O}:$) and the <u>triple</u> bond in each molecule of N_2 ($:N{:}{:}{:}N:$ or $:N{\equiv}N:$).

Since electrons are "shared" -- not transferred -- in covalent bonds, the "octet rule", which states that atoms tend to either gain or lose electrons until they have <u>eight</u> valence electrons each, needs to be modified somewhat in order to predict what the electrons in covalent molecules will do.

$$\text{THE MODIFIED OCTET RULE:}$$

The <u>modified octet rule</u> states that the atoms of the "main-group" elements tend to gain or lose electrons until they each have <u>eight</u> electrons in their valence shells, <u>even if some of those eight electrons are "shared" with other atoms</u>. Consider the Lewis structures of Cl_2, HCN, and CO_2 shown below:

$$:\!\overset{..}{Cl}\!:\!\overset{..}{Cl}\!:\ \text{or}\ :\!\overset{..}{Cl}\!-\!\overset{..}{Cl}\!:\qquad H\!:\!C\!:\!:\!:\!N\!:\ \text{or}\ H\!-\!C\!\equiv\!N\!:\qquad :\!\overset{..}{O}\!:\!:\!C\!:\!:\!\overset{..}{O}\!:\ \text{or}\ :\!\overset{..}{O}\!=\!C\!=\!\overset{..}{O}\!:$$

Notice that each atom (except H) in the above Lewis structures is surrounded by a total of <u>eight</u> electrons, whether they are used in covalent bonds or not. (Pairs of electrons which are <u>not</u> part of a covalent bond are usually called <u>lone pairs</u>.) In determining this, notice that electrons in <u>bonds</u> are counted <u>twice</u> -- once for each atom. For example, each chlorine atom in Cl_2 is surrounded by <u>eight</u> valence electrons, despite the fact that there are only <u>fourteen</u> valence electrons present in the entire molecule! The two electrons in the <u>bond</u> are part of <u>both</u> "octets".

Notice that we said "except H" above. Hydrogen is an <u>exception</u> to the octet rule; hydrogen atoms only require <u>two</u> electrons in each of their valence shells. (This should make sense: the valence shell of a hydrogen atom is the <u>first</u> shell, and the first shell ($n = 1$) can hold a maximum of <u>two</u> electrons.) Other exceptions include elements in Group IIA (beryllium (Be) only requires <u>four</u> valence electrons) and Group IIIA (boron (B) only requires <u>six</u> valence electrons). In addition, elements which lie <u>below the second row</u> of the Periodic Table <u>may</u> include <u>more</u> than eight electrons in their valence shells if necessary. For example, the sulfur atom in the SF_6 molecule is surrounded by <u>twelve</u> electrons.

In general, the number of covalent bonds connected to the atoms of a "main-group" element is (8 - GN), where "GN" is the group number of the element. Thus, <u>carbon</u> atoms have <u>four</u> bonds, <u>nitrogen</u> atoms have <u>three</u>, <u>oxygen</u> atoms have <u>two</u>, and <u>halogen</u> atoms have one. (Check the Lewis structures above!)

HOW TO DRAW LEWIS STRUCTURES:

The correct Lewis structures for many molecules and polyatomic ions can be easily drawn simply by following the rules outlined below.

Step # 1: Count the total number of <u>valence electrons</u> for each atom in the molecule. Add these numbers together, <u>add one</u> for each <u>negative</u> charge present on the molecule or ion, and <u>subtract one</u> for each <u>positive</u> charge present.

Step # 2: Draw the <u>skeletal structure</u> (the basic arrangement of the atoms) of the molecule or ion. If there is a <u>unique atom</u> (that is, only <u>one</u> atom of a particular element), place that atom in the <u>center</u> of the molecule or ion.

Step # 3: Connect adjacent atoms with <u>single bonds</u>.

Step # 4: Surround all atoms <u>except</u> the central atom with enough lone pairs to complete their octets. (Remember the exceptions to the octet rule!)

Step # 5: Count the number of valence electrons in your drawing. (Each single bond or lone pair = <u>two</u> electrons.) Subtract the total from the total you got in Step # 1. If any valence electrons remain, place them on the <u>central</u> atom in the form of lone pairs (whenever possible).

Step # 6: If the central atom does not have a complete "octet" of electrons, complete its "octet" by moving lone pairs on adjacent atoms toward the central atom, forming <u>multiple bonds</u> until the central atom's "octet" is complete.

Step # 7: Check for violations of the "8 - GN" rule, which predicts the number of covalent bonds that should be attached to each atom.

The following is the result of each step above for a molecule of <u>HNO_2</u>.

Step # 1: H = 1, N = 5, O = 6, O = 6. 1 + 5 + 6 + 6 = <u>18</u> electrons.

Step # 2: H O N O (<u>N</u> is the central atom, since H can only have <u>one</u> bond.)

Step # 3: H-O-N-O   Step # 4: H-Ö-N-Ö: Step # 5: H-Ö-N-Ö: (18 - 16 = 2 e⁻'s.)

Steps # 6 and # 7: H-Ö-N=Ö: (H-O=N-O: violates the "8 - GN" rule.)

DRAWING LEWIS STRUCTURES:

Problem: Draw the correct Lewis structure for the ClF_3 molecule.

Solution: Cl = 7, F = 7.   7 + 7 + 7 + 7 = <u>28</u> valence electrons.
Place the unique atom (Cl) in the center, then surround it with three fluorine
atoms.   Filling in the single bonds (three) and lone pairs around the fluorine
atoms (three lone pairs each, for a total of nine) requires the use of a total of
<u>24</u> electrons.   28 - 24 = <u>4</u> electrons which must be placed on the central Cl atom.
The final Lewis structure is shown at the right.   Note that the Cl
atom is surrounded by <u>ten</u> electrons.   This violation of the octet
rule is permissible here, since Cl is below the second row of the Periodic Table.

Problem: Draw the correct Lewis structure for the CS_2 molecule.

Solution: C = 4, S = 6.   4 + 6 + 6 = <u>16</u> valence electrons.   Place the
unique atom (C) in the center, then place the sulfur atoms on each side.   Filling
in the single bonds (two) and lone pairs around the sulfur atoms (three lone pairs
each, for a total of six) requires the use of a total of <u>16</u> electrons.   No
electrons remain to be added to the central atom (16 - 16 = 0).   However, the
carbon atom is only surrounded by <u>four</u> electrons -- it needs four more to complete
its octet.   This problem is solved by moving one lone pair from each sulfur atom
toward the central carbon atom, forming <u>two double bonds</u>:   :S=C=S:

Problem: Draw the correct Lewis structure for the <u>cyanide ion (CN⁻)</u>.

Solution: C = 4, N = 5, negative charge = 1.   4 + 5 + 1 = <u>10</u> valence
electrons.   Simply connect the two atoms with a single bond, then complete the
octet of either atom with three lone pairs.   This requires <u>eight</u> electrons, which
leaves <u>two</u> electrons to go onto the other atom (10 - 8 = 2).   At this point, one
atom will only have <u>four</u> electrons surrounding it.   To complete this atom's octet,
move <u>two</u> lone pairs from the other atom to form a <u>triple bond</u>:   :C≡N:

RESONANCE STRUCTURES:

Problem: Draw the correct Lewis structure for ozone (O_3).

Solution: Oxygen is in Group VIA. $3 \times 6 = \underline{18}$ valence electrons. Since there is no unique atom here, place any atom in the center and surround it with the other two -- that is, arrange the three oxygen atoms in a straight line. Filling in the single bonds (two) and lone pairs around the outer oxygen atoms (three lone pairs each, for a total of six) requires the use of a total of 16 electrons. $18 - 16 = \underline{2}$ electrons which must be placed on the central oxygen atom. This leaves us with the structure :O-O-O:, in which the central oxygen atom only has six valence electrons surrounding it. To complete the octet, one lone pair should be moved in from one of the outer oxygen atoms to form a double bond. It makes no difference which of the two outer oxygen atoms donates a lone pair to the central atom, since they are essentially equivalent. Therefore, the final Lewis structure of ozone could be written as either :O=O-O: or as :O-O=O: -- either one is a reasonable possibility. When more than one reasonable Lewis structure can be drawn for a molecule, the structures are called resonance structures. Ozone does not really "resonate" between these two structures. The actual structure of ozone is a hybrid of the two structures. This is symbolized as shown below.

:O=O-O: ⟷ :O-O=O: = :O===O===O: (actual structure)

The double-headed arrow above is used to indicate resonance structures. The dotted line between the oxygen atoms in the hybrid structure indicates that the bond it represents is not a full covalent bond, but a sort of "half-bond" -- the two electrons in question are distributed over all three oxygen atoms. This may sound strange, but it is important to realize that the actual structure of ozone is best represented by the resonance hybrid structure above. The resonance structures just help us to visualize the actual structure a little more easily.

FORMAL CHARGES AND COORDINATE COVALENT BONDS:

An atom which carries an electrical charge will violate the "8 - GN" rule, which is used to predict the number of covalent bonds normally attached to that atom. Electrical charges which are located on atoms are sometimes called formal charges. For an example of this, consider the hydronium ion (H_3O^+), for which the Lewis structure is drawn below:

$$H\!:\!\overset{..}{\underset{..}{O}}\!: \quad + \quad H^+ \quad \longrightarrow \quad H\!:\!\overset{..}{\underset{..}{O}}\!:\!H$$
$$\quad\quad H \quad\quad\quad\quad\quad\quad\quad\quad\quad H$$

Since oxygen is in Group IIA, a normal oxygen atom should only have two covalent bonds. (8 - 6 = 2.) The oxygen atom in the hydronium ion has three covalent bonds. This is acceptable as long as we realize that the oxygen atom in the hydronium ion also carries the ion's positive electrical charge. Another example is the BH_4^- ion, for which the Lewis structure is shown below:

$$\begin{array}{ccc} H & & H \\ H\!:\!B & + \quad :\!H^- \quad \longrightarrow \quad & H\!:\!B\!:\!H \\ H & & H \end{array}$$

Boron is an exception to the "8 - GN" rule. Boron is in Group IIIA, and its atoms normally have three covalent bonds. In the BH_4^- ion, the boron atom has four covalent bonds. Again, this indicates that the boron atom is the atom that carries the ion's negative electrical charge.

The polyatomic ions above each contain a coordinate covalent bond, which is just like any other covalent bond except for the fact that the two electrons in the bond came originally from the same atom or ion. A simple way to determine the formal charge on an atom is to use the equation: F.C. $= GN - U - \dfrac{S}{2}$, where GN is the atom's group number, U is the number of unshared electrons around the atom, and S is the number of shared electrons surrounding the atom.

VALENCE SHELL ELECTRON PAIR REPULSION THEORY:

Electrons, being negatively charged, tend to <u>repel</u> each other. This property of electrons is the basis of the <u>Valence Shell Electron Pair Repulsion</u> <u>(VSEPR) Theory</u>, which states that the <u>electron pairs</u> in the <u>valence shell</u> of an atom will tend to arrange themselves in such a way as to be <u>as far apart from each other as possible</u>. The shapes of covalent molecules can be predicted from their Lewis structures by applying the VSEPR theory to the <u>central</u> atom in the molecule. In this context, "electron pairs" refers to either <u>lone pairs</u> or <u>covalent bonds</u>, regardless of whether the covalent bonds are single bonds or multiple bonds.

An atom with <u>two</u> pairs of electrons in its valence shell would have, as its lowest energy state, the electron pairs located on <u>opposite sides</u> of the atom. This is called a <u>linear</u> geometry, since the electron pairs and the atom all lie on one line. If <u>three</u> electron pairs are present in an atom's valence shell, they will be arranged in a <u>triangular</u> pattern around the atom. <u>Four</u> electron pairs produce a <u>tetrahedral</u> geometry. (A <u>tetrahedron</u> is a four-sided figure which looks like a pyramid with a triangular base. If the atom is located at the center of a tetrahedron, the electron pairs would point toward its corners.) <u>Five</u> pairs of electrons produce a <u>trigonal bipyramidal</u> geometry. (A <u>trigonal bipyramid</u> is a six-sided figure which looks like two pyramids stuck together at one triangular face. If the atom is located at the center of a trigonal bipyramid, the electron pairs would point toward its corners.) <u>Six</u> pairs of electrons produce a geometry called <u>octahedral</u>. (An <u>octahedron</u> is an eight-sided figure which looks like two square-based pyramids stuck together at the square base. If the atom is located at the center of an octahedron, the electron pairs would point toward its corners.) These five basic geometries are sufficient to describe the shapes of a large number of covalent molecules.

PREDICTING THE SHAPES OF MOLECULES USING VSEPR THEORY:

Problem: What is the geometry of the CO_2 molecule?

Solution: The Lewis structure of CO_2 is shown at the right. $:\overset{..}{O}=C=\overset{..}{O}:$

The carbon atom is the central atom, and it is surrounded by only two "electron pairs" (each double bond counts as only one "electron pair" when using the VSEPR theory).   The molecule is linear, as indicated in the Lewis structure.   Notice that this geometry implies an O-C-O bond angle of 180°.

Problem: What is the geometry of the BH_3 molecule?

Solution: The Lewis structure of BH_3 is shown at the right.   H:B:H

The boron atom is the central atom, and it is surrounded by only three electron pairs (three single bonds), so the molecule has a triangular geometry.   A better drawing of the BH_3 molecule is shown at the right.   Notice that this geometry implies three equal H-B-H bond angles of 120°.

Problem: What is the geometry of the CH_4 molecule?

Solution: The Lewis structure of CH_4 is shown at the right.
From the Lewis structure, a square geometry with 90° H-C-H bond angles would be predicted, but the actual geometry of the CH_4 molecule is the tetrahedral geometry, with H-C-H bond angles of 109.5^O.   One feature of the VSEPR theory is that the bond angles should always be as large as possible.

Problem: What is the geometry of the H_2O molecule?

Solution: The Lewis structure of H_2O is shown at the right.   H:O:

Like CH_4, H_2O has four electron pairs surrounding the central atom.   Unlike CH_4, two of those electron pairs are lone pairs.   The lone pairs repel the single bonds, giving the H_2O molecule a bent geometry (not linear like CO_2!).   The H-O-H bond angle in H_2O is approximately 105°, due to the fact that lone pairs repel other electron pairs a little bit more strongly than covalent bonds do.

A SUMMARY OF MOLECULAR GEOMETRIES:

The table below summarizes the terms used to describe the geometries of molecules with various numbers of electron pairs surrounding the central atom.

Total Electron Pairs Around Central Atom	Total Lone Pairs Around Central Atom	Term Used To Describe Geometry Of Molecule	One Example Of A Molecule With This Geometry
2	0	Linear	CO_2
	1	Linear	N_2
3	0	Triangular	BH_3
	1	Bent	SO_2
4	0	Tetrahedral	CH_4
	1	Pyramidal	NH_3
	2	Bent	H_2O
	3	Linear	HF
5	0	Trigonal Bipyramidal	PCl_5
	1	"Seesaw"-Shaped	SF_4
	2	"T"-Shaped	ClF_3
	3	Linear	XeF_2
6	0	Octahedral	SF_6
	1	Square Pyramidal	BrF_5
	2	Square	XeF_4

Bond angles for each of the five basic geometries:

Linear	180°
Triangular	120°
Tetrahedral	109.5°
Trigonal Bipyramidal	90° and 120°
Octahedral	90°

POLAR COVALENT BONDS AND ELECTRONEGATIVITY:

In some covalent bonds, the electrons are distributed symmetrically between the connected atoms. In other covalent bonds, this is not the case -- one atom has a greater share of the electrons in the bond than the other atom does. If the electrons in a covalent bond are more attracted to one atom in the bond than to the other atom, then the atom which has the greater share of the electrons will also have a partial negative charge. The atom which has the lesser share of the electrons will be left with a partial positive charge. The covalent bond thus has two "opposite ends", or <u>poles</u>. It is therefore called a <u>polar covalent bond</u>. The partial positive and partial negative charges in a polar covalent bond are represented by "$\delta+$" and "$\delta-$" respectively. ("δ" is the lower-case Greek letter <u>delta</u>.) These charges are not as strong as the full positive ("+") and negative ("-") charges present in ionic bonds, but they are measurable.

An example of a molecule which contains a polar covalent bond is HF. The fluorine atom draws the electrons in the covalent bond toward itself and thus acquires a $\delta-$ charge, leaving the hydrogen atom with a $\delta+$ charge. Atoms such as fluorine, which tend to draw the electrons in covalent bonds toward themselves, are called <u>electronegative</u>. Fluorine atoms are the most electronegative, oxygen atoms are second, nitrogen atoms are third, and chlorine and bromine fourth and fifth. Examples of molecules which contain <u>non-polar</u> covalent bonds are H_2 and F_2. In these cases, the connected atoms are <u>equal</u> in electronegativity (obviously -- they're identical atoms!), so the electrons in the bond are arranged symmetrically, and no partial electrical charges are present.

The polarity of a molecule is a <u>vector</u> quantity -- it has a numerical value, and it <u>points in a certain direction</u>. Polar covalent bonds are often represented by arrows which point to the "$\delta-$" atom, as shown for HF: H-F

POLARITY OF MOLECULES:

Molecules such as HF, which contain only one polar covalent bond, can easily be seen to be polar. When molecules have more than one polar covalent bond, we must consider the polarity of the <u>individual</u> bonds and the <u>directions</u> in which their polarity vectors point in order to be able to determine whether the molecule as a whole is polar or nonpolar. A few examples are shown below.

<u>Problem</u>: Are molecules of H_2O polar?

<u>Solution</u>: The Lewis structure of H_2O is shown at the right. The molecule is <u>bent</u>, due to the presence of the lone pairs on the oxygen atom. As shown in the drawing, the polarity vectors point in different directions. The sum of the two polarity vectors is the large vector shown above. (To add two vectors, draw one vector so that its "tail" is at the same point as the "head" of the other vector. The sum is the new vector which goes from the "tail" of the second vector to the "head" of the first vector.) This vector indicates that H_2O molecules have a positive end and a negative end, and are therefore <u>polar</u> molecules.

<u>Problem</u>: Are molecules of CO_2 polar?

<u>Solution</u>: The Lewis structure of CO_2 is shown at the right. Notice that while each C=O bond is itself polar, the two polarity vectors point in exactly opposite directions. Adding these two vectors produces a sum of <u>zero</u>. Therefore, molecules of CO_2 are <u>nonpolar</u>.

<u>Problem</u>: Are molecules of SO_2 polar?

<u>Solution</u>: The Lewis structure of SO_2 is shown at the right. (Actually, this is just one of the two possible resonance structures.) The lone pair on the sulfur atom imparts a <u>bent</u> geometry to the SO_2 molecule. The polarity vectors do <u>not</u> cancel each other out. SO_2 is <u>polar</u>.

BOND ENERGIES AND BOND LENGTHS:

When two atoms are brought together, the valence electrons of each atom are attracted to the nucleus of the other atom. This results in the "sharing" of electrons which we call a covalent bond. When two objects which attract each other are brought closer together, the potential energy of the objects is lowered. Thus, the formation of a covalent bond is an energy-lowering (or exothermic) process. Consider the formation of a chlorine molecule from two chlorine atoms, shown below:

$$:\!\overset{..}{\underset{..}{Cl}}\!\cdot \; + \; \cdot\!\overset{..}{\underset{..}{Cl}}\!: \; \longrightarrow \; :\!\overset{..}{\underset{..}{Cl}}\!:\!\overset{..}{\underset{..}{Cl}}\!: \qquad \Delta H^0_r = -58.066 \text{ kcal/mole}^1$$

Not surprisingly, the reverse of this process -- the breaking of a covalent bond to produce two electrically neutral fragments -- is an energy-consuming (endothermic) process. To separate one mole of chlorine molecules into two moles of chlorine atoms, 58.066 kilocalories of energy must be supplied.

The amount of energy needed to break a covalent bond into electrically neutral fragments is called the bond energy of that covalent bond. Multiple bonds generally have larger bond energies than single bonds. For example:

Bond	Bond Energy, kcal/mole[1]	Bond Length, pm[1]
C-C	88	154.1
C=C	172	133.7
C≡C	230	120.4

(Note: 1 picometer (pm) = 10^{-12} meters.)

The bond length for a covalent bond is simply the distance between the atoms in that bond when the molecule is in its lowest energy state. (If the two atoms are too close together, their nuclei will repel each other, and if they are too far apart, the attraction of the valence electrons for the nuclei will pull them together.) Single bonds are generally longer than multiple bonds.

---

1) CRC Handbook of Chemistry and Physics, 56th Edition, 1975-1976.

AVERAGE BOND ENERGIES:

Any given covalent bond will have approximately the same bond energy regardless of the compound in which it appears. For example, a C-H bond will have a bond energy of approximately 100 kcal/mole, whether that C-H bond is one of the bonds in CH_4, C_3H_8, CH_2O, or any other compound. Numerous calorimetry experiments have allowed scientists to determine the <u>average bond energy</u> of many different kinds of covalent bonds. These average bond energies allow scientists to estimate the heat of a reaction <u>without</u> doing additional calorimetry experiments.

<u>Problem</u>: Estimate the value of ΔH_r^0 for the reaction shown below:

$$\text{H-C}\equiv\text{C-H} \; + \; :\overset{\shortmid\shortmid}{\text{Cl}}\text{-}\overset{\shortmid\shortmid}{\text{Cl}}: \longrightarrow :\overset{\shortmid\shortmid}{\text{Cl}}\text{-}\underset{\text{H}}{\overset{\shortmid\shortmid}{\text{C}}}=\underset{\text{H}}{\overset{\shortmid\shortmid}{\text{C}}}\text{-}\overset{\shortmid\shortmid}{\text{Cl}}:$$

<u>Solution</u>: Lewis structures have been drawn above to make it easier to see that in the course of the reaction, the C≡C bond and the Cl-Cl bond are <u>broken</u> (they are present in the reactants, but not in the products), and the C=C bond and the two C-Cl bonds are <u>formed</u> (they are present in the products, but not in the reactants). <u>Bond-breaking</u> is an <u>energy-consuming (endothermic)</u> process, and <u>bond-making</u> is an <u>energy-releasing (exothermic)</u> process. Hence, if we know the average bond energies, we can estimate the amount of energy released or consumed by the overall reaction. For the bonds above, the average bond energies[1] are:

C≡C = 230 kcal/mole   C=C = 172 kcal/mole

Cl-Cl = 58 kcal/mole   C-Cl = 85 kcal/mole

Now, a little arithmetic gives us the value of ΔH_r^0! ("ABE" = average bond energy)

$$H_r^0 = ABE_{(broken)} - ABE_{(formed)}$$
$$= ABE_{C\equiv C} + ABE_{Cl-Cl} - ABE_{C=C} - ABE_{C-Cl} - ABE_{C-Cl}$$
$$= 230 + 58 - 172 - 85 - 85 = \underline{-54 \text{ kcal/mole}}.$$

---

1) CRC Handbook of Chemistry and Physics, 56th Edition, 1975-1976.

VALENCE BOND THEORY:

Atoms of <u>carbon</u> contain a total of <u>six</u> electrons. The electron configuration of a carbon atom is represented by the energy-level diagram at the right. Notice that there are <u>two</u> unpaired electrons in the 2p subshell of a carbon atom. It would be reasonable to expect carbon atoms to form <u>two</u> covalent bonds by sharing those electrons with <u>two</u> other atoms having one unpaired electron each -- for example, hydrogen atoms. The compound formed when hydrogen and carbon react should have the formula CH_2 by the above reasoning. In fact, molecules of CH_2 are very unstable. A much more stable molecule is CH_4 (methane). Methane molecules have a <u>tetrahedral</u> geometry -- all of the H-C-H bond angles are 109.5°. This is also difficult to explain using the above logic -- p orbitals are arranged at 90° angles to each other. Obviously, a better theory is needed to explain the bonding in CH_4 and similar compounds.

The <u>valence bond theory</u> was proposed by Heitler and London in 1927. It says that since electrons in atoms are located in <u>orbitals</u>, and since electrons are "shared" between two atoms in a covalent bond, then covalent bonds can only be made by bringing orbitals together so that they overlap. The greater the overlap of the atomic orbitals, the stronger the covalent bond. However, atomic orbitals are just the <u>graphs</u> of the Schrödinger <u>wave equations</u>. They can therefore be treated just like any other mathematical functions -- they can be added, subtracted, multiplied, etc. When these operations are performed on mathematical functions, new functions result. Similarly, when atomic orbitals from the same atom are mathematically combined, <u>new orbitals</u> result. These new orbitals are called <u>hybrid orbitals</u>. Hybrid orbitals resemble the orbitals from which they are made. For example, an s orbital and a p orbital can be "hybridized" to give <u>two sp hybrid orbitals</u>, which are roughly spherical (like s orbitals) but have two lobes each (like p orbitals).

HYBRID ORBITALS:

The following rules should be obeyed whenever it is necessary to combine atomic orbitals into hybrid orbitals to explain the bonding in a molecule.

Rule # 1: The orbitals being hybridized must all come from the same atom. Orbitals from different atoms are never hybridized together.

Rule # 2: Only orbitals in the valence shell of an atom may be hybridized. (This should make sense, since valence electrons form covalent bonds.)

Rule # 3: Orbitals that are relatively low in energy are hybridized before orbitals that are relatively high in energy. s orbitals and p orbitals are hybridized before d orbitals and f orbitals.

Rule # 4: Any number of orbitals may be hybridized. However, the number of hybrid orbitals obtained at the end of the hybridization process must be the same as the number of atomic orbitals used at the beginning of the process. For example, an s orbital and a p orbital (a total of two orbitals) may be combined to form a set of two "sp" hybrid orbitals.

Rule # 5: The hybrid orbitals formed by the hybridization process will arrange themselves around their atom so as to be as far apart from each other as possible. For example, the two sp hybrid orbitals formed in the example in Rule 4, above, will point in exactly opposite directions. The 180^0 angle between them implies a linear geometry for the covalent bonds attached to that particular atom.

The geometries of other sets of common hybrid orbitals are given below.

sp^2 hybrid orbitals: three orbitals, triangular geometry, 120^0 angles.

sp^3 hybrid orbitals: four orbitals, tetrahedral geometry, 109^0 angles.

sp^3d hybrid orbitals: five orbitals, trigonal bipyramidal, 90^0 & 120^0.

sp^3d^2 hybrid orbitals: six orbitals, octahedral geometry, 90^0 angles.

(Note: "sp^2" is pronounced "S-P-two", not "S-P-squared".)

HOW TO DETERMINE THE HYBRIDIZATION OF AN ATOM:

Problem: What kind of hybrid orbitals are used by carbon in CH_4 ?

Solution: If you look at the Lewis structure at the right, you should see that the carbon atom in CH_4 is surrounded by a total of four pairs of electrons (four covalent bonds). Each pair of electrons _must_ be located in an _orbital_, so a set of _four_ orbitals is needed by the carbon atom in order to bond with four hydrogen atoms. A set of $\underline{sp^3}$ hybrid orbitals is made from an s orbital and three p orbitals -- a total of _four_ orbitals. Thus, the set of sp^3 hybrid orbitals must contain four orbitals, which is just the number the carbon atom needs. Therefore, the carbon atom in CH_4 is said to be $\underline{sp^3\text{-hybridized}}$. Since sp^3 hybrid orbitals have a _tetrahedral_ geometry, we can safely predict that the CH_4 molecule will be tetrahedral as well.

Problem: What kind of hybrid orbitals are used by chlorine in ClF_3 ?

Solution: See the Lewis structure at the right. The Cl atom is surrounded by a total of _five_ electron pairs (two lone pairs and three covalent bonds). Therefore, a set of _five_ hybrid orbitals is needed, which means that _five_ atomic orbitals must have been used to make the set of hybrid orbitals. The five orbitals of lowest energy in the valence shell of a chlorine atom are the _s_ orbital, the three _p_ orbitals, and one of the five _d_ orbitals. s + p + p + p + d = a set of $\underline{sp^3d}$ hybrid orbitals. Cl is $\underline{sp^3d\text{-hybridized}}$ in ClF_3.

Problem: What are the hybridizations of carbon and oxygen in CO_2 ?

Solution: Multiple bonds are counted as only _one_ pair of electrons when determining hybridizations. Therefore, the carbon atom has _two_ pairs of electrons around it (two double bonds), so it is _sp-hybridized_. Each oxygen atom has _three_ electron pairs around it (two lone pairs and the double bond), so each oxygen atom is $\underline{sp^2\text{-hybridized}}$.

The reason that multiple bonds are counted as <u>one</u> pair of electrons (just like single bonds and lone pairs) when determining the hybridization of an atom is that multiple bonds are composed of two different types of covalent bonds. One type is made from the <u>hybrid</u> orbitals surrounding the atom, and the other is made from the "leftover" (unhybridized) atomic orbitals. Each multiple bond contains only <u>one</u> of the type of bond made from hybrid orbitals, so it is counted only as a single bond -- what we're <u>really</u> counting are the hybrid orbitals!

The two types of covalent bonds are <u>sigma</u> (lower-case Greek letter σ) bonds and <u>pi</u> (lower-case Greek letter π) bonds. In a <u>sigma</u> bond, the electrons which make up the bond are most likely to be found <u>directly between</u> the connected atoms. In a <u>pi</u> bond, the electrons which make up the bond are most likely to be found <u>above</u>, <u>below</u>, or <u>alongside</u> the connected atoms.

<u>Single</u> bonds, which are formed by the "end-to-end" overlapping of atomic orbitals, are simple examples of <u>sigma</u> bonds. <u>Multiple</u> bonds also contain <u>sigma</u> bonds, but they also contain one or more <u>pi</u> bonds. The <u>pi</u> bonds are formed by the "sideways" overlapping of the atomic orbitals which remain after the hybridization process is complete. Specifically, a <u>single</u> bond is made of one <u>sigma</u> bond, a <u>double</u> bond is made of one <u>sigma</u> bond and one <u>pi</u> bond, and a <u>triple</u> bond is made of one <u>sigma</u> bond and <u>two pi</u> bonds. (To help you visualize this, consider a hot dog in a bun. The hot dog looks something like a sigma bond -- the "meat" lies directly between the two ends of the hot dog. The bun looks something like a pi bond -- it lies above and below the hot dog, but not directly between the two ends of the hot dog. It is also important to note that <u>one</u> pi bond consists of <u>two</u> regions in space where electrons may be found, just as a hot dog bun consists of <u>two</u> pieces of bread -- the top piece and the bottom piece.)

HOW PI BONDS ARE FORMED:

The valence bond theory states that covalent bonds are formed whenever atomic orbitals overlap. The difference between sigma bonds and pi bonds is that <u>sigma</u> bonds are formed by the <u>"end-to-end"</u> overlapping of <u>hybrid</u> orbitals, while <u>pi</u> bonds are formed by the <u>"side-to-side"</u> overlapping of <u>unhybridized</u> orbitals.

Consider an atom which is $\underline{sp^2}$-hybridized. Since the valence shells of most atoms contain <u>one</u> s orbital and <u>three</u> p orbitals, a <u>p</u> orbital will be "left over" after a set of three sp^2 hybrid orbitals has been made from the s orbital and two of the three p orbitals. The three sp^2 hybrid orbitals lie in the same plane, arranged in a triangular fashion (120° apart). The "leftover" p orbital is perpendicular to the sp^2 hybrid orbitals, projecting above and below the plane which contains them. If two sp^2-hybridized atoms are brought together so that one sp^2 hybrid orbital from one atom overlaps with one sp^2 hybrid orbital from the other atom in an "end-to-end" fashion, the two overlapping sp^2 hybrid orbitals form a sigma bond. However, a <u>pi</u> bond will be formed at the same time, since the two "leftover" p orbitals on the two atoms will overlap in a "side-to-side" fashion. The result is a <u>double bond</u> -- one sigma bond and one pi bond.

<u>Triple bonds</u> are formed in a similar way, using <u>sp</u>-hybridized atoms. The two sp hybrid orbitals are arranged in a linear fashion (180° apart). There are <u>two</u> "leftover" p orbitals -- one projects above and below the line on which the two sp hybrid orbitals lie, and the other projects in front of and behind the two sp hybrid orbitals. A sigma bond is formed when two sp-hybridized atoms are brought together so that an sp hybrid orbital on one atom overlaps with an sp orbital on the other atom. Simultaneously, the p orbitals which project above and below the sigma bond overlap to form one pi bond, and the p orbitals which project in front of and behind the sigma bond overlap to form the other pi bond.

GASES AND THEIR PROPERTIES:

Most of the matter on earth occurs in one of three physical <u>states</u>, or <u>phases</u> -- namely, the <u>solid</u> state, the <u>liquid</u> state, or the <u>gaseous</u> state. <u>Gases</u> obey some very simple physical laws, which makes it somewhat easier to study gases as a group than to study liquids and solids (the <u>condensed</u> phases).

When you blow up a balloon, the air inside the balloon has the same shape as the balloon does. Because gases tend to take on the same shape as their containers, they are classified as <u>fluids</u>. The <u>volume</u> of a sample of a gas is just the volume of its container. The <u>pressure</u> of a gas is simply the ratio of the force exerted by the gas particles (atoms or molecules) against the walls of their container to the surface area of those walls. (Pressure of any kind is defined as force/area.) Hence, gas pressure is often measured in <u>pounds per square inch</u>. (Atmospheric pressure is roughly 14.7 p.s.i. at sea level.)

In 1643, the Italian physicist Evangelista Torricelli showed that air exerts pressure by filling a long glass tube which had been sealed at one end with mercury, and then inverting the tube into a container of mercury which was open to the atmosphere. Some of the mercury moved down into the container, but most of it remained in the glass tube, held in place by the pressure of the air! Torricelli found that the height of the mercury column in the glass tube was approximately 760 millimeters above the surface of the mercury in the container. Hence, another common unit used to measure pressure is <u>mmHg</u> ("millimeters of mercury") or <u>inHg</u> ("inches of mercury"). "Milliliters of mercury" are sometimes called <u>Torr</u> in honor of Torricelli. The most common unit used by scientists to measure pressure is the <u>atmosphere</u> (<u>atm</u>). One standard atmosphere = 760 mmHg (760 Torr) = 14.7 p.s.i. = 29.92 inHg. ("Inches of mercury" are used by meteorologists to report the barometric pressure. Torricelli's apparatus is called a <u>barometer</u>.)

BOYLE'S LAW:

In 1666, the Irish chemist Robert Boyle performed the first quantitative experiments on gases. Boyle used a sealed glass tube similar to the one Torricelli used, but bent into the shape of the letter "J", with the sealed end at the short stem of the "J". By pouring mercury into the open end of the tube, Boyle found that he was able to "trap" a small of air in the sealed end of the tube. Furthermore, the volume of this trapped sample of air <u>decreased</u> when more mercury was added to the tube -- that is, when the pressure on the sample of air <u>increased</u>. Careful measurements of the pressure and volume of the gas sample at several different pressures and volumes led Boyle to the conclusion that volume and pressure are <u>inversely proportional</u> -- when one goes up by a certain amount, the other goes down by the same amount. Boyle's law can be written as:

$$PV = constant$$

where P is the pressure and V is the volume of the sample of gas. Since the product of the pressure and the volume is a constant, another way to write this is:

$$P_i V_i = P_f V_f$$

where the subscript "i" refers to the <u>initial</u> values of pressure and volume for the gas sample, and the subscript "f" refers to their <u>final</u> values.

<u>Problem</u>: A balloon has a volume of 2.54 L at 1.00 atm pressure. If it is taken to a higher altitude where the pressure is only 0.750 atm, what will be the volume of the balloon?

<u>Solution</u>: Rewriting the second equation above, we get:

$$V_f = \frac{P_i V_i}{P_f}$$

Now, just substitute in the correct values of P_i, V_i, and P_f:

$$V_f = \frac{(1.00 \text{ atm}) \times (2.54 \text{ L})}{(0.750 \text{ atm})} = 3.39 \text{ L}.$$

<u>Note</u>: The balloon <u>increases</u> in volume, as it should at the lower pressure.

90

CHARLES' LAW:

In 1787, the French physicist Jacques Charles measured the _volumes_ of samples of various gases at different _temperatures_. He plotted his data on a graph of volume vs. temperature, and noted that the data points for each gas lay on a straight line. The lines for different gases had different slopes.

Charles drew the following conclusions from these results:

1) The _volume_ of a sample of a gas is _directly proportional_ to the _temperature_ of the gas. As the temperature _increases_, the volume also _increases_. In equation form, Charles' law can be written as:

$$V = (\text{constant}) \times T$$

where V is the volume and T is the _Kelvin_ temperature of the sample of the gas.

2) The lines on the graph all converged at the point on the graph which corresponded to a volume of _zero_ and a temperature of _-273_ $^{\circ}$C. This is the lowest possible temperature, called _absolute zero_ (zero on the _Kelvin_ scale).

Problem: A balloon has a volume of 750 mL in a freezer whose temperature is -10 $^{\circ}$C. If the balloon is removed from the freezer and allowed to warm to room temperature (25 $^{\circ}$C), what will be its volume?

Solution: A convenient form of Charles' law to use in this case is:

$$\frac{V_i}{T_i} = \frac{V_f}{T_f}$$

where "i" and "f" indicate the _initial_ and _final_ conditions, respectively. Using the above temperatures as written leads to a _negative_ value for the new volume. This isn't physically possible -- there's no such thing as a "negative volume"! Converting the Celsius temperatures into _Kelvin_ temperatures solves this problem:

$$-10 + 273 = 263 \text{ K} = T_i \qquad 25 + 273 = 298 \text{ K} = T_f$$

$$V_f = \frac{V_i T_f}{T_i} = \frac{(750 \text{ mL}) \times (298 \text{ K})}{(263 \text{ K})} = 850 \text{ mL}. \quad (\underline{\text{Note}}: \text{ The balloon } \underline{\text{expands}}.)$$

THE COMBINED GAS LAW:

Boyle's law and Charles' law can be combined into one equation. This equation is sometimes called the combined gas law, and it is shown below:

$$\frac{PV}{T} = (constant)$$

where P is the pressure of the gas, V is the volume of the gas, and T is the Kelvin temperature of the gas. Another way to write this equation is:

$$\frac{P_i V_i}{T_i} = \frac{P_f V_f}{T_f}$$

where "i" and "f" refer to the initial and final conditions, respectively. Notice that if the temperature is kept constant (that is, if $T_i = T_f$), then this equation is simply the equation for Boyle's law ($P_i V_i = P_f V_f$). Similarly, if the pressure is kept constant ($P_i = P_f$), then this equation becomes the Charles' law equation.

Problem: A balloon has a volume of 810 mL at a pressure of 750 Torr and a temperature of -77 ^{o}C. (This is the temperature of a mixture of "dry ice" (frozen CO_2) and acetone.) The balloon is removed from the dry ice/acetone bath and allowed to warm to room temperature (25 ^{o}C) in a decompression chamber in which the pressure is 3.00 atmospheres. Calculate the final volume of the balloon.

Solution: This problem is difficult to solve intuitively -- warming the balloon should cause it to expand, but increasing the pressure should cause it to contract. However, this problem is easily solved using the combined gas law. We need to remember to use only Kelvin temperatures (no negative numbers allowed!) and to be consistent in all other units (760 Torr = 1.00 atm). Hence:

$$-77 + 273 = 196 \text{ K} = T_i \qquad 25 + 273 = 298 \text{ K} = T_f$$

$$3.00 \text{ atmospheres} \times 760 \text{ Torr/atm} = 2280 \text{ Torr} = P_f$$

$$V_f = \frac{P_i V_i T_f}{T_i P_f} = \frac{(750 \text{ Torr}) \times (810 \text{ mL}) \times (298 \text{ K})}{(196 \text{ K}) \times (2280 \text{ Torr})} = 405 \text{ mL}.$$

92

AVOGADRO'S LAW:

In 1808, the French chemist Joseph Gay-Lussac discovered that when gases react with each other, the _volumes_ of the gases which react are related to each other by _small whole-number ratios_. For example, when hydrogen gas reacts with oxygen gas to form water vapor, _two_ liters of hydrogen are consumed for each liter of oxygen that is consumed, and _two_ liters of water vapor are formed in this process. The point is that the ratios _by volume_ of hydrogen, oxygen, and water vapor involved are 2:1:2 -- small whole numbers. (This principle is sometimes called the _Law of Combining Volumes_.)

At about this same time, the English scientist John Dalton proposed his atomic theory, which stated that atoms combine in small whole-number ratios to form molecules during a chemical reaction. In 1811, the Italian chemist Amadeo Avogadro showed that the law of combining volumes and the atomic theory would both make sense if _equal volumes of different gases contained equal numbers of particles (atoms or molecules) of the gases_. For example, in the case above, _two_ molecules of H_2 react with _one_ molecule of O_2 to form _two_ molecules of H_2O -- the ratio is 2:1:2, whether we're talking about _volumes_ of gases (in liters) or _amounts_ of gases (in numbers of molecules). The underlined principle above is sometimes called _Avogadro's law_, and can be written in equation form as follows:

$$V = (\text{constant}) \times n$$

where V is the volume of the gas and n is the amount of gas present, in _moles_ (the usual units used for counting small things like molecules!). This shouldn't be too surprising -- if you want to increase the volume of a balloon, one way to do it is to blow more air into it! However, it is important to realize that the volume of the balloon depends on the _amount_ of gas present (in _moles_), not on the mass of the gas or the atomic weight or molecular weight of the gas particles.

$$\text{THE IDEAL GAS LAW:}$$

The equations for Boyle's law, Charles' law, and Avogadro's law can be combined into one equation, as shown below. This equation is called the <u>ideal gas law</u>, and any gas which obeys this law is called an <u>ideal gas</u>.

$$\text{Boyle:} \quad PV = \text{constant} \qquad \text{Charles:} \quad \frac{V}{T} = \text{constant} \qquad \text{Avogadro:} \quad \frac{V}{n} = \text{constant}$$

$$\text{Ideal Gas Law:} \quad \frac{PV}{nT} = \underline{\text{universal gas constant}} = \underline{R} = 0.082056 \, \frac{\text{L atm}}{\text{K mole}}.$$

(It should be noted that most <u>real</u> gases do not obey the ideal gas law <u>exactly</u>, but many of them come reasonably close to ideal behavior under most conditions.) The ideal gas law is usually written as $\underline{PV = nRT}$. The other gas laws can be obtained from the ideal gas law by simply keeping all quantities constant except the two quantities being studied. For example, if pressure and temperature are kept constant, $PV = nRT$ simply becomes $V = n \times (\text{constant})$, which is Avogadro's law. The <u>units</u> of the gas constant (R) may look strange, but remembering them is useful, since the correct units for each quantity must be used when using $PV = nRT$.

<u>Problem</u>: What volume is occupied by 1.00 mole of an ideal gas at a temperature of 0 $^\circ$C and a pressure of 1.00 atmosphere?

<u>Solution</u>: Rearranging $PV = nRT$ gives us:

$$V = \frac{nRT}{P}$$

For the units to cancel properly, the temperature must be in <u>Kelvins</u>:

$$T = 0 + 273 = 273 \text{ K}$$

$$V = \frac{(1.00 \text{ mole}) \times (0.082056 \, \frac{\text{L atm}}{\text{K mole}}) \times (273 \text{ K})}{(1.00 \text{ atm})} = 22.4 \text{ L.}$$

The temperature and pressure given above are referred to as the <u>standard temperature and pressure</u>, or <u>STP</u> for short. The volume calculated above is called the <u>standard molar volume</u> of an ideal gas -- that is, the volume occupied by one mole of an ideal gas at standard temperature and pressure.

$$\text{USING THE IDEAL GAS LAW:}$$

Problem: A 50.00-mL sealed glass tube contains 88.0 mg of an unknown gas at 27.0 $^{\circ}$C and 750 Torr. What is the molecular weight of the gas?

Solution: Rearranging PV = nRT allows us to find the amount of gas present (in moles). Remembering to use the correct units gives us the following:

$$n = \frac{PV}{RT} \qquad P = 750 \text{ Torr} \times \frac{1 \text{ atm}}{760 \text{ Torr}} = 0.987 \text{ atm}$$

$$V = 50.00 \text{ mL} = 50.00 \times 10^{-3} \text{ L}$$

$$T = 27.0 + 273 = 300 \text{ K}$$

$$n = \frac{(0.987 \text{ atm}) \times (50.00 \times 10^{-3} \text{ L})}{(0.082056 \frac{\text{L atm}}{\text{K mole}}) \times (300 \text{ K})} = 2.00 \times 10^{-3} \text{ moles}$$

Molecular weight has units of g/mole, so divide the mass of the gas by this value:

$$88.0 \text{ mg} = 88.0 \times 10^{-3} \text{ grams}$$

$$\text{Molecular Weight} = \frac{88.0 \times 10^{-3} \text{ grams}}{2.00 \times 10^{-3} \text{ moles}} = 44.0 \text{ g/mole.}$$

This is an example of the Dumas method for determining the molecular weight of a gas. One possible identity for the gas is CO_2 (12 + 16 + 16 = 44).

Problem: Calcium carbonate decomposes upon heating. The reaction is:

$$CaCO_{3\,(s)} \longrightarrow CaO_{(s)} + CO_{2\,(g)}$$

What volume of CO_2 at STP is produced when 304 grams of $CaCO_3$ is decomposed?

Solution: The molecular weight of $CaCO_3$ is 100.09 g/mole. Therefore:

$$(304 \text{ g } CaCO_3) \times \frac{1 \text{ mole } CaCO_3}{100.09 \text{ g } CaCO_3} \times \frac{1 \text{ mole } CO_2}{1 \text{ mole } CaCO_3} = 3.04 \text{ moles } CO_2$$

The ideal gas law could be used at this point, but since we're working at STP, we can simply use the standard molar volume of a gas: 22.4 L = 1 mole of gas at STP.

$$3.04 \text{ moles } CO_2 \times \frac{22.4 \text{ L}}{1 \text{ mole}} = 68.0 \text{ L.}$$

Note: This "trick" only works at standard temperature and pressure! At other temperatures and pressures, the ideal gas law must be used.

DALTON'S LAW OF PARTIAL PRESSURES:

In the early part of the 19th century, the English scientist John Dalton (the same John Dalton who proposed the atomic theory!) was studying mixtures of gases. Not surprisingly, he found that each gas in a mixture of gases expands to fill its container. (This should make sense -- consider the air in the room where you are now. Air is a mixture of nitrogen, oxygen, and other gases, all of which expand to fill the room. Anywhere you go in the room, you can find oxygen to breathe -- fortunately!) Therefore, the pressure exerted by one gas in a mixture of gases is the same as it would be if the gas were the only gas present in the container. Dalton called the pressures of the individual gases in the gas mixture partial pressures. Furthermore, he determined that the total pressure exerted by a mixture of gases is equal to the sum of the partial pressures of the individual gases in the mixture. In equation form, this can be written as:

$$P_m = P_1 + P_2 + P_3 + P_4 + \ldots$$

where P_m is the total pressure of the mixture of gases and P_1, P_2, P_3, P_4, and so on are the partial pressures exerted by gas # 1, gas # 2, gas # 3, gas # 4, etc.

One way to prepare samples of pure gases in the laboratory is to do a chemical reaction which produces the desired gas and pass the gas from the vessel in which the reaction is occurring into a bottle filled with water. The gas displaces the water, leaving the pure gas in the bottle. However, this gas isn't really "pure" -- it's mixed with water vapor! As a liquid evaporates, the vapors that are produced behave like any other gas. For instance, the molecules of the vapors collide with the walls of their container, exerting a pressure which is called the vapor pressure of the liquid from which they came. As the temperature of a liquid rises, its vapor pressure rises. The liquid's boiling point is the temperature at which its vapor pressure equals the atmospheric pressure.

96

USING DALTON'S LAW OF PARTIAL PRESSURES:

Problem: Potassium chlorate decomposes on heating. The reaction is:

$$2\ KClO_3\ (s) \longrightarrow 2\ KCl\ (s) + 3\ O_2\ (g)$$

The oxygen formed by heating a sample of $KClO_3$ was collected by passing it into a 500-mL flask filled with water at a temperature of 30 ^{0}C. The oxygen displaced all of the water in the flask. The atmospheric pressure was 764 Torr. The vapor pressure of water at 30 ^{0}C is 32 Torr. What mass of $KClO_3$ was used?

Solution: Since a _mixture_ of gases (oxygen and water vapor) is obtained in this experiment, we must correct the pressure for the presence of the undesired gas (water vapor). Using Dalton's law of partial pressures, we get:

$$P_m = P_{oxygen} + P_{water\ vapor}$$

$$P_{oxygen} = P_m - P_{water\ vapor} = 764\ Torr - 32\ Torr = 732\ Torr$$

Now, use the ideal gas law to find the amount of oxygen present (in moles). Be sure to use the correct _units_ for all quantities, as shown below:

$$P_{oxygen} = 732\ Torr \times \frac{1\ atm}{760\ Torr} = 0.963\ atm$$

$$V_{oxygen} = 500\ mL = 0.500\ L$$

$$T_{oxygen} = T_{water} = 30 + 273 = 303\ K$$

$$PV = nRT$$

$$n = \frac{PV}{RT} = \frac{(0.963\ atm) \times (0.500\ L)}{(0.082056\ \frac{L\ atm}{K\ mole}) \times (303\ K)} = 0.0194\ moles\ O_2$$

Now, using the balanced equation above, we can calculate the amount of $KClO_3$ that decomposed (in moles). Conversion of this to a mass is straightforward, as shown:

$$0.0194\ moles\ O_2 \times \frac{2\ moles\ KClO_3}{3\ moles\ O_2} = 0.0129\ moles\ of\ KClO_3$$

$$0.0129\ moles\ KClO_3 \times \frac{122.553\ g\ KClO_3}{1\ mole\ of\ KClO_3} = 1.58\ grams\ of\ KClO_3$$

GRAHAM'S LAW OF EFFUSION:

If you've ever owned a helium-filled balloon, you've probably noticed that helium-filled balloons deflate more rapidly than air-filled balloons do. In either case, the balloon deflates because the gas particles (atoms or molecules) inside the balloon gradually escape from the balloon through small pores in the walls of the balloon. This process is called effusion. Effusion techniques were used during World War II to separate the isotopes of uranium.

In 1829, the Scottish chemist Thomas Graham discovered that the rate at which a gas undergoes effusion is inversely proportional to the square root of the molecular weight of the gas. In equation form, this can be written as:

$$\text{Rate of Effusion} = \frac{\text{constant}}{(\text{M.W. of gas})^{1/2}}$$

This can also be written as follows when comparing the effusion rates of two gases:

$$\frac{\text{Rate of Effusion, Gas \# 1}}{\text{Rate of Effusion, Gas \# 2}} = \frac{(\text{M.W. of Gas \# 2})^{1/2}}{(\text{M.W. of Gas \# 1})^{1/2}}$$

Problem: Which undergoes effusion at a faster rate -- He or CH_4 ? Calculate the relative rates of effusion of these two gases.

Solution: Use the above equation. Let's make He = Gas # 1:

$$\frac{\text{Rate of Effusion of He}}{\text{Rate of Effusion of CH}_4} = \frac{(\text{M.W. of CH}_4)^{1/2}}{(\text{M.W. of He})^{1/2}} = \frac{(16.043 \text{ g/mole})^{1/2}}{(4.00260 \text{ g/mole})^{1/2}}$$

$$= 4.0054/2.0006 = 2.0020$$

Thus, it can be seen that helium undergoes effusion about twice as rapidly as CH_4 does. This should make sense -- helium atoms are lighter (less massive) than CH_4 molecules. Therefore, if a helium atom and a methane molecule have the same amount of kinetic energy (K.E. = $mv^2/2$), the less massive particle (helium) should have the greater velocity. The relative rates of effusion of two gases are determined by the relative velocities with which their particles move.

THE KINETIC-MOLECULAR THEORY OF GASES:

In the 1860's, the Scottish physicist James Clerk Maxwell and the German physicist Ludwig Boltzmann developed a model to try to explain why gases behave as they do. This model is now called the <u>kinetic-molecular theory of gases</u>. The essential ideas of the kinetic-molecular theory are outlined below:

1. Ideal gases consist of particles which are so small (compared with the relatively large distances between them) that they can be considered to have essentially <u>zero</u> volume. Gases have volumes because of the relatively large distances between the gas particles, not because of the size of the particles.

2. The particles of an ideal gas do not attract or repel each other. However, they do move around constantly, and occasionally may collide with each other and with the walls of their container. Gases have pressures due to the force exerted by the collisions of their particles with the walls of their containers.

3. During a collision between two particles, kinetic energy may be transferred from one particle to another. Hence, two different particles of the <u>same</u> sample of a gas may have <u>different</u> kinetic energies. The <u>average</u> kinetic energy of a sample of gas particles is directly proportional to the <u>temperature</u> of the gas, in <u>Kelvins</u>.

This model can be used to explain many of the properties of gases. For example, gases are <u>compressible</u>, but liquids and solids are not compressible. This is because compressing a gas simply forces the particles of the gas closer together. The particles in liquids and solids are already close together, so that they cannot be compressed. Also, why should there be a "coldest possible temperature" (absolute zero) ? The answer is that as the temperature decreases, the kinetic energy of the gas molecules decreases, causing their <u>velocity</u> to decrease. Absolute zero is the temperature at which the molecules <u>stop</u> moving!

THE KINETIC-MOLECULAR THEORY AND THE FUNDAMENTAL GAS LAWS:

The principles of the kinetic-molecular theory of gases can be used to explain many of the fundamental gas laws. Examples are given below.

Boyle's Law: The pressure exerted by a gas comes from the force with which the gas particles strike the walls of their container. If the volume of the container increases, the surface area of the inner walls of the container increases as well. If the number of particles present and the temperature remain constant, the total force exerted against the walls of the container remains the same, so the pressure (the average force per unit of surface area) decreases.

Charles' Law: As the temperature of a gas increases, the kinetic energy of the gas particles increases. Since K.E. = $mv^2/2$, and since the mass (m) of the gas particles doesn't change, the result is that the velocity (v) of the gas particles increases. When the rapidly-moving gas particles collide with the walls of the container, they exert a greater force on the walls than slowly-moving gas particles do. Hence, the pressure of the gas sample will increase. However, if the walls of the container are movable (for example, the flexible rubber in a balloon), the increased force pushes them out, so the volume of the gas increases.

Graham's Law: The rate of effusion of a gas is directly proportional to the velocity of the gas particles. Since K.E. = $mv^2/2$, the velocity of the gas particles can be expressed as: $v = [(2) \times (K.E.)/m]^{1/2}$. If the temperature of the gas remains constant, the value of K.E. will remain constant. Therefore, the rate of effusion of the gas will be directly proportional to $m^{-1/2}$, where m is the mass of the gas particles -- that is, the molecular weight of the gas!

Similar arguments can be made for Avogadro's Law and for Dalton's Law. This is another illustration of the principal work of the scientist -- to construct a model which accurately predicts and/or explains the observations made in nature!

LIQUIDS AND SOLIDS:

The behavior of _gases_ can be described by relatively simple models such as the kinetic-molecular theory because the individual gas particles act independently of each other. The attractive forces between the gas particles are relatively small, due to the relatively large distances between them. However, in the _condensed_ phases (_liquids_ and _solids_), the distances between the particles are much smaller, and the attractive forces between the particles are therefore much greater. Because of this, the particles do _not_ act independently of each other. Therefore, no simple model will accurately describe the behavior of liquids and solids. However, some generalizations can be made about their behavior.

Unlike gases, liquids and solids cannot be compressed. (An important feature of the braking system of most cars is the transmission of force from the brake pedal to the brake shoes through the _brake fluid_ (a liquid). If liquids were compressible, this force could not be transmitted, and the brakes wouldn't work.) Unlike solids, liquids are _fluids_ -- they take on the shapes of their containers, and they have flow rates which can be measured. (Gases are also fluids.) A unique feature of liquids is their _surface tension_, which is caused by the attraction of liquid particles below the liquid surface for liquid particles at the surface.

A _phase diagram_ is a graph of pressure vs. temperature which summarizes the conditions under which a substance exists as a solid, as a liquid, or as a gas. Not surprisingly, most substances are _solids_ at _high pressures_ and _low temperatures_ and _gases_ at _high temperatures_ and _low pressures_, with liquids being favored at intermediate conditions of pressure and temperature. At the boundary lines between the solid, liquid, and gas regions of a phase diagram, more than one phase can exist. At the point where all three regions meet on the phase diagram (called the _triple point_), all three phases exist simultaneously.

PHASE TRANSITIONS AND DYNAMIC EQUILIBRIUM:

When a substance undergoes a change from one physical state to another, the process is referred to as a phase transition. Common examples of phase transitions include the melting of ice (solid to liquid), the freezing of water (liquid to solid), the boiling of water (liquid to gas), and the condensation of water to form dew (gas to liquid). The process in which a solid is converted into a gas directly (without passing through a liquid phase) is called sublimation. (Solids which have odors, such as solid air fresheners, sublime to a small degree.)

Phase transitions depend upon the kinetic energies of the particles involved and the attractive forces between the particles. For example, evaporation occurs when liquid particles with relatively high kinetic energy approach the liquid surface and enter the gas phase by overcoming the attractive forces in the liquid phase. Similarly, condensation occurs when gas particles strike a liquid or solid surface and transfer some of their kinetic energy to that surface. The particles no longer have enough kinetic energy to overcome the attractive forces in the condensed phase, so they remain in the liquid or solid phase.

Now, consider a butane cigarette lighter. Butane (C_4H_{10}, b.p. 0 $^{\circ}$C) is normally a gas at room temperature, but the lighter contains both liquid butane and gaseous butane in contact with each other. Butane molecules continually move back and forth between the liquid phase and the gas phase. Thus, both vaporization and condensation (two opposing phase transitions) are occurring simultaneously. The system as a whole does not change, but individual molecules are constantly changing phases. This situation is one example of a dynamic equilibrium, and is written as:

$$C_4H_{10} \text{ (l)} \rightleftharpoons C_4H_{10} \text{ (g)}$$

The arrows pointing in each direction indicate that two opposing phase transitions are occurring simultaneously. The system is dynamic (changing) but at equilibrium.

LeCHATELIER'S PRINCIPLE:

Consider the following equilibrium, which occurs in a butane lighter:

$$C_4H_{10}\ (l) \ \rightleftharpoons\ C_4H_{10}\ (g)$$

When the lighter is ignited, it is the butane <u>gas</u> (nearest the top of the lighter) which reacts with oxygen in the air to produce the lighter flame. Therefore, the amount of gaseous butane in the lighter decreases. This decreases the likelihood that a molecule of butane in the gas phase will strike the liquid butane surface and re-enter the liquid phase. Therefore, the two phase transitions above (liquid to gas and gas to liquid) are no longer equally probable, and the system is not at equilibrium any more. However, the system can easily re-establish the equilibrium if some of the liquid butane evaporates to provide a fresh supply of butane gas. In fact, this is what happens -- the lighter works until the <u>liquid</u> butane is gone.

This is a specific example of a more general principle that applies to systems in a state of dynamic equilibrium. Dynamic equilibrium is a <u>low-energy</u> state, and any system will try to minimize its energy whenever possible. Thus, if a system in a state of dynamic equilibrium is disturbed in such a way so that it is no longer at equilibrium, then the system will readjust itself (if possible) to try to counteract the disturbance and restore the equilibrium. This principle was first stated in 1888 by the French physical chemist Henri Louis LeChatelier, and is known as <u>LeChatelier's Principle</u>.

In the above case, the "disturbance" is the ignition of the lighter. This consumes C_4H_{10} (g) and removes it from the system. "Disturbances" in the sense of LeChatelier's Principle generally change the amount of material on one side of the equation. The "readjustment" is the liquid-to-gas phase transition of butane, restoring the gaseous butane which was removed by the "disturbance". When the liquid butane runs out, this "adjustment" is no longer possible.

HYDROGEN BONDING:

If the intermolecular attractive forces which hold liquid molecules in the liquid phase are relatively _strong_, then a greater amount of kinetic energy will be needed for an individual molecule to overcome these forces and enter the gas phase. The greater the _temperature_ of the molecules, the greater their kinetic energy. At high temperatures, the vapor pressure of a liquid is greater than at low temperatures, since at high temperatures, more liquid molecules have enough kinetic energy to overcome the intermolecular attractive forces and enter the gas phase. Therefore, a compound's _boiling point_ (the temperature at which its vapor pressure equals 1.00 atmosphere) can be useful in determining the strength of the intermolecular attractive forces which hold its molecules in the liquid phase.

Consider water (H_2O, b.p. 100 ^{O}C) and hydrogen sulfide (H_2S, b.p. -60 ^{O}C). Since higher temperatures are needed to cause water to boil than to cause hydrogen sulfide to boil, we can conclude that the intermolecular attractive forces between water molecules are stronger than the intermolecular attractive forces between hydrogen sulfide molecules. This is due to the fact that water molecules are much more _polar_ than hydrogen sulfide molecules. Oxygen atoms are much more electronegative than sulfur atoms, and will therefore carry a stronger partial negative charge than sulfur atoms. This means that the hydrogen atoms in water molecules will carry a stronger partial positive charge than the hydrogen atoms in hydrogen sulfide molecules. The strong partial positive charges on the hydrogen atoms in water molecules will be attracted to the strong partial negative charges on the oxygen atoms in other water molecules. This causes water molecules to be more attracted to each other than hydrogen sulfide molecules are. The attraction of a partially positive hydrogen atom for a partially negative atom in another molecule is called _hydrogen bonding_ (a relatively _strong_ intermolecular force).

104

INTERMOLECULAR ATTRACTIVE FORCES:

Compare the boiling points of the compounds below:

```
      H H H H              H H H H              H H H H              HHH
      " " " "              " " " "              " " " "              " " "
     :O:C:C:O:            :O:C:C:C:H          H:C:C:C:C:H          H C H
      " "                  " " "               " " " "             " " "
      H H                  H H H               H H H H           H:C:C:C:H
                                                                  " " "
                                                                  H H H
```

   Ethylene Glycol   Propanol   Butane   Isobutane

   b.p. 198 ^{o}C   b.p. 97 ^{o}C   b.p. 0 ^{o}C   b.p. -12 ^{o}C

Notice that the boiling points decrease as the amount of hydrogen bonding (that is,
the number of O-H bonds present) decreases. (Ethylene glycol (anti-freeze) has _two_
O-H bonds, propanol has _one_ O-H bond, and butane and isobutane have _no_ O-H bonds.)
Hydrogen bonding is one example of a class of intermolecular attractive forces
called <u>dipole-dipole interactions</u>. Dipole-dipole interactions occur in molecules
that have polar bonds, and consist simply of the attraction between a partially
positive atom in one molecule and a partially negative atom in another molecule.
This is not restricted to bonds involving hydrogen atoms; for example, molecules of
CO_2 are held together in the solid state ("dry ice" is solid CO_2) by means of
dipole-dipole interactions. (CO_2 is a <u>nonpolar</u> molecule but contains polar bonds.)

   <u>London forces</u> are the primary intermolecular attractive forces between
molecules which do <u>not</u> contain polar bonds. These forces come about because the
electrons in atoms and molecules are constantly in motion. When a majority of the
electrons in a molecule move to one side of the molecule, that side of the molecule
takes on a partial negative charge, leaving a partial positive charge on the other
side of the molecule. These partial charges make up an <u>instantaneous dipole</u> -- it
only exists for an <u>instant</u> before the electrons move again. London forces are just
the attractions between instantaneous dipoles in different molecules. Elongated
molecules (such as butane) have a larger surface area than more "spherical"
molecules (such as isobutane), and thus have greater attractive London forces.

CRYSTALLINE SOLIDS:

Many solids are made of particles which are arranged in highly ordered, repeating patterns called _lattices_. Lattices are usually represented by their _unit cells_, which are small segments of the lattice in question. Repeating the pattern of the unit cell many times in all directions generates the entire lattice. The simplest types of lattices are those based on _cubic_ unit cells. The simplest cubic unit cell is simply eight particles arranged at the corners of a cube. This is called a _simple cubic_ unit cell. Variations on this theme include the arrangement which includes a particle at the center of the cube (_body-centered cubic_), and the arrangement which includes a particle at the center of each of the six faces of the cube (_face-centered cubic_). Other unit cells are variations of the simple cubic structures in which one side of the "cube" is longer than the others or shorter than the others, or in which some of the angles between the sides of the "cube" are greater than 90° or less than 90°.

Solids whose particles are arranged in lattices are called _crystalline_ solids (or just _crystals_). Crystalline solids can usually be classified as one of four basic types. _Ionic_ crystals (such as NaCl) are composed of cations and anions which are held in their lattice sites by ionic bonds. They are hard and brittle, with high melting points. They don't conduct electricity well when solid, but the _solutions_ made by dissolving them in water conduct electricity. _Covalent_ crystals (such as diamonds) are composed of atoms held together by covalent bonds. They are hard crystals with high melting points, and generally don't conduct electricity. _Molecular_ crystals (such as ice) are composed of molecules held together by dipole-dipole attractions or London forces. They are soft crystals with low melting points. _Metallic_ crystals (_metals_, such as iron) are composed of cations held in place by a "sea" of electrons around them. Metals conduct electricity very well.

PROPERTIES OF CRYSTALLINE AND AMORPHOUS SOLIDS:

In an ionic crystal, the cations and anions are next to each other.
This represents a very stable arrangement for these ions, since positive charges
and negative charges attract each other. (This is called <u>Coulomb's Law</u>.) Many of
the properties of ionic crystals can be explained on this basis. For example,
ionic crystals generally have high melting points. (NaCl melts at over 800 ^{o}C.)
This is due to the fact that the attractive forces between ions are very strong,
so a large amount of energy (that is, heat) must be supplied to overcome these
forces and melt the crystal. The brittleness of ionic crystals is due to the fact
that ions of similar charge are forced together when the crystal is subjected to
extreme pressure. These ions tend to <u>repel</u> each other, causing the crystal to
break apart. Also, electric currents flow when charged particles (such as ions or
electrons) are free to move. Ions trapped in a rigid crystalline lattice are <u>not</u>
free to move, but become free when the crystal is melted or dissolved in water.
Thus, <u>solutions</u> of ions conduct electricity, but ionic <u>crystals</u> do not.

The melting of a crystal occurs at a <u>constant temperature</u> if the
pressure remains constant. (For example, ice melts at 0 ^{o}C at 1.00 atmosphere.)
This is called a <u>first-order phase transition</u>. However, some substances (such as
glass and plastics) do not undergo first-order (constant temperature) phase
transitions, but instead gradually soften as they are heated. Solids which behave
in this way are called <u>amorphous solids</u>. ("Amorphous" means "without shape".) The
molecules of amorphous solids are usually long, "spaghetti"-like strands of atoms
which do not fit conveniently into a lattice arrangement. Hence, amorphous solids
are sometimes called <u>non-crystalline solids</u>. Since "crystallization" never really
occurs for an amorphous solid, it is sometimes called a <u>supercooled liquid</u>. (The
<u>supercooling</u> process cools a liquid below its freezing point <u>without</u> its freezing.)

APPENDIX -- CHEMICAL NOMENCLATURE:

Chemical <u>nomenclature</u> is just the system by which chemical compounds are <u>named</u>. The system for ionic compounds is different from the system for covalent compounds. Both systems are described below.

<u>Ionic</u> compounds are named by either of <u>two</u> methods. In the first method, the cation is named first, and the anion is named with the suffix "-ide". Thus, KBr is <u>potassium bromide</u>, and CaO is <u>calcium oxide</u>. When transition metals (which form more than one cation) are used, the <u>charge</u> on the cation is included in Roman numerals after the cation's name. Thus, CuCl is copper(I) chloride, and $CuCl_2$ is copper(II) chloride. The other method that is sometimes used for naming transition metal compounds involves using the metal's <u>Latin</u> name, with a suffix attached to distinguish between the possible cations. Under this system, CuCl is <u>cuprous</u> chloride, whereas $CuCl_2$ is <u>cupric</u> chloride. (The suffix "-ic" indicates the cation with the <u>larger</u> charge, while the suffix "-ous" indicates the cation with the <u>smaller</u> charge.) It is important to keep the two systems separate: FeO is either <u>iron(II) oxide</u> or <u>ferrous oxide</u>, whereas Fe_2O_3 is either <u>ferric oxide</u> or <u>iron(III) oxide</u>, but not "ironic oxide"!

<u>Covalent</u> compounds are named as though they were ionic compounds, with the "cation" first and the "anion" last. Greek prefixes are used to indicate the number of each atom present in the compound. The prefixes used are "mono-" (1), "di-" (2), "tri-" (3), "tetra-" (4), "penta-" (5), "hexa-" (6), "hepta-" (7), "octa-" (8), "nona-" (9), "deca-" (10), and so on. Thus, CO_2 is <u>carbon dioxide</u>, CCl_4 is <u>carbon tetrachloride</u>, CO is <u>carbon monoxide</u>, and N_2O_5 is <u>dinitrogen pentoxide</u>. (Note that in the cases of "monoxide" and "pentoxide", the prefix has been shortened slightly so as to make pronunciation easier.) This system should <u>not</u> be used with ionic compounds -- Fe_2O_3 is <u>not</u> "diiron trioxide"!

APPENDIX -- NOMENCLATURE OF ACIDS AND THEIR SALTS:

The rules for naming binary acids and their salts are different from the rules for naming oxyacids and their salts. Binary acids (those composed of hydrogen and one other element) are named by placing the prefix "hydro-" and the suffix "-ic acid" around the name (sometimes abbreviated) of the "other" element. Thus, HCl is hydrochloric acid, and HI is hydriodic acid. The salts of binary acids have names which end in "-ide". For example, NaCl is sodium chloride, and KI is potassium iodide.

The names of oxyacids (those composed of hydrogen, oxygen, and one other element) are found using a somewhat more complex system of rules. Prefixes and suffixes are still used, but the particular combination of a prefix and a suffix is what provides information about the number of oxygen atoms present in the molecules of the acid. This situation arises because any given element may be able to form several oxyacids. If only two oxyacids are known for a given element, then prefixes are not used, and the suffixes "-ic acid" and "-ous acid" are used to indicate the larger number of oxygen atoms and the smaller number of oxygen atoms, respectively. Thus, HNO_3 is nitric acid, whereas HNO_2 is nitrous acid. If more than two oxyacids are possible for a given element, then the prefixes "hypo-" and "per-" are used to indicate the minimum number of oxygen atoms and the maximum number of oxygen atoms, respectively. Thus, $HClO_4$ is perchloric acid, $HClO_3$ is chloric acid, $HClO_2$ is chlorous acid, and $HClO$ is hypochlorous acid. The salts of oxyacids whose names end in "-ic acid" have names which end in "-ate", while the salts of oxyacids whose names end in "-ous acid" have names which end in "-ite". Thus, KNO_3 is potassium nitrate, and $NaClO$ is sodium hypochlorite. Salts which still contain acidic hydrogen atoms are named using a prefix to indicate the number of hydrogen atoms present. Thus, NaH_2PO_4 is sodium dihydrogen phosphate.